James Davies

Galvanized Iron

its Manufacture and Uses - A Detailed Description of this Important Industry and

its Manufacturing Process

James Davies

Galvanized Iron

its Manufacture and Uses - A Detailed Description of this Important Industry and its Manufacturing Process

ISBN/EAN: 9783337106362

Printed in Europe, USA, Canada, Australia, Japan

Cover: Foto ©berggeist007 / pixelio.de

More available books at **www.hansebooks.com**

GALVANIZED IRON

Its Manufacture and Uses

A DETAILED DESCRIPTION
OF THIS IMPORTANT INDUSTRY AND ITS
MANUFACTURING PROCESS.

BY JAMES DAVIES

London
E. & F. N. SPON, LTD., 125 STRAND

New York
SPON & CHAMBERLAIN, 12 CORTLANDT STREET

1899

PREFACE.

I COMMENCED this Treatise several years since, but laid it on one side in consequence of other engagements. Having been asked several times for such particulars as I have embraced in my work, I determined to bring it to a conclusion and publish it.

I do not remember having hitherto seen any publication dealing with this important industry to the extent it deserves, and I therefore feel every confidence that my labour will be appreciated. I have not written theoretically, but from a practical point of view.

It is the result of a connection with the trade of over thirty years—first as manager (for my father, the late Edward Davies, who was the third to commence in the industry), and afterwards as managing partner, of one of the most flourishing works at that time in the Midland Counties. It will be observed that the prices given are not the prices of to-day, as it is obvious prices might even change while the book is being printed. For this reason I give the prices current at the time I commenced the work.

JAMES DAVIES.

82 REDPOST LANE,
FOREST GATE, LONDON, E.:
September 1899.

CONTENTS.

Importance of the Galvanized Iron Trade—Comparative Rates of Carriage—Advantages of South Wales—Cost of Materials . . . *pages* 1 *to* 8

How to succeed in the Business—How to Fail—Arrangement of a Works—Requirements of the various Markets—The best Markets to cultivate *pages* 9 *to* 42

Methods of Packing and Prices—Brands—Corrugating of Galvanized Iron—Rolling from Steel Bars *pages* 42 *to* 56

Heathfield's Patent Process—Bayliss' Patent Process—Original Process—The Best Process . *pages* 57 *to* 68

Weight of Coating—Close Annealing — Pickling — Labour Prices—Prices of Galvanized Sheets—Best and Best Best—Extras . . . *pages* 68 *to* 82

Patent Flattened Sheets—Black Corrugated Sheets—Continuous Roofing—Roofing Tiles—Galvanized Tinned Sheets . . . . . *pages* 83 *to* 91

To ascertain the Quality of Spelter — Galvanizing Baths—Stopping Leaky Baths — Treatment of Zinc Ashes—Flux Skimmings . . . *pages* 91 *to* 102

Treatment of Dross—Curved and Ridged Roofs—
Girth of Curved Roofs—Approximate Cost of
Curved Roofs — To Estimate the Weight of a
Roof—Weights of Galvanized Sheets . *pages* 103 *to* 116

Prices for Fixing—Girths of Ridged Roofs—Cost of
making Gutters and Ridging . . . *pages* 117 *to* 125

Galvanizing Abroad—Approximate Cost of Machinery
and Plant for producing Galvanized Corrugated
Iron Ridging, Gutters, Tanks and Cisterns,
Buckets, &c., also Plant for Galvanizing same
pages 125 *to* 136

INDEX . . . . *pages* 137 *to* 139

GALVANIZED IRON.

IMPORTANCE OF THE GALVANIZED IRON TRADE.

THE importance and volume of this trade can readily be ascertained from a consideration of the following particulars.

The total shipments for 1894 amounted to nearly 170,000 (one hundred and seventy thousand) tons, an advance on 1893 of 3200 tons, and on 1892 of nearly 13,000 tons.

Australia was the best customer, taking 41,700 (forty-one thousand seven hundred) tons; India next, with 28,000 (twenty-eight thousand) tons; South Africa, 20,600 (twenty thousand six hundred) tons; and the Argentine, 18,000 (eighteen thousand) tons.

Alike, as regards Australia, the Argentine and South Africa, the demand was larger than in either of the two previous years, and, as regards India, the demand was almost equal to 1893 or 1892.

The total value for 1894 was £1,952,066 (one million nine hundred and fifty-two thousand and sixty-six pounds!).

In 1893 the value of the shipments was £2,045,770; in 1892, £2,077,068.

The total for three years was the enormous sum of *six millions and seventy-four thousand nine hundred and four pounds.*

The remarkable manner in which the export demand for galvanized iron kept up and increased in 1895 is a matter of much gratifying comment; and *last year has beaten all previous records*, and surpassed even the greatest expectations of the makers, who were reckoning on a total shipment of 200,000 tons at the most. Instead, the figures come out 204,200 (two hundred and four thousand two hundred) tons. So that last year *exceeded* the previous year by 34,200 (thirty-four thousand two hundred) tons. This is a marvellous performance for this class of iron.

The largest customer on the year has been Australia, with 40,780 (forty thousand seven hundred and eighty) tons. India, which has been rapidly improving for twelve months past, takes a second place, with 38,800 (thirty-eight thousand eight hundred) tons. South Africa, which has advanced as rapidly as India, has been the third largest buyer, having purchased 33,500 (thirty-three thousand five hundred) tons. The Argentine Republic is fourth, with 19,600 (nineteen thousand six hundred) tons. "Other countries," unenumerated, appear for 20,600 (twenty thousand six hundred) tons, and Chili has taken 10,800 (ten thousand eight hundred) tons. *Never before have such returns been presented to the galvanized iron trade, and their condition affords the liveliest prospects for the year now entered upon!*

The value of the exports for the last four years, 1892–95, is the enormous amount of over

EIGHT AND A-HALF MILLIONS STERLING,

and the first three months of this year show an increase of twenty-five per cent. in the volume of the exports as against the corresponding period of 1895.

These returns for four years show that this trade has been increasing by "leaps and bounds," and forms a very striking contrast to the tin-plate industry. One of the chief reasons for this remarkable increase is the comparative ease with which galvanized corrugated sheets can be fixed by unskilled labour, and the small comparative quantity of timber required in the construction of the framework of the roof. If timber is not available iron principals and purlins answer equally as well, as the sheets can be attached by clips or roof-hooks. It therefore follows that the development of fresh countries, and the increase in agricultural pursuits, leads the way for the introduction of this favourite mode of roofing. Unlike tin-plates, it has no competition to fear from the countries to which it is exported, as the conditions of manufacture are such as to render it impossible to carry on this industry abroad and compete successfully with the works here. Several works have started with this object in view, but the scarcity of the requisite labour, the climate, the difficulty of getting regular supplies of materials, and, above all, the heavy capital required to cover the supplies of material afloat, in course of manufacture, and the finished article in stock, have invariably led to the collapse of such undertakings.

I have now shown that the dimensions of this trade are not only immense, but increasing yearly. At the same time, the margin of profit now, unlike former years, is within such a compass that it can only be attained by working the business on systematic lines, and gaining experience from the success or failures of others, as I will proceed to explain in detail hereafter.

There is nothing to prevent any firm, possessing the necessary capital, energy and intelligence, from attaining to a high standard of success in this business.

HALF A MILLION

of money was accumulated in twenty-five years by a member of this trade, now deceased, whose will was recently proved, and this indicates that there is money to be made in the business.

COMPARATIVE RATES OF CARRIAGE.

As a basis I have taken Swansea as being the manufacturing centre, and any difference in carriage for other places can easily be ascertained.

The following are the railway rates from Swansea, and also from the Midlands, to the following chief ports:—

RATES FOR GALVANIZED IRON
(AT OWNERS' RISK).

TO	From Wolverhampton			From Swansea		
	10-ton lots.	4-ton loads.	2-ton loads.	10-ton lots.	4-ton loads.	2-ton loads.
	s. d.	s. d.	s. d.	s. d.	s. d.	s. d.
Liverpool (G.W.R.)	10 0	..	..	12 6	..	..
London (G.W.R.)	12 6	..	..	..	12 6	..
Glasgow	..	17 6	..	..	25 0	..
Hull	12 6	..	..	..	19 2	..
Bristol (G.W.R.)	..	..	10 0	..	6 8	..
Southampton (L.S.W.R.)	12 6	..	..	..	..	12 6
Newport, Mon. (G.W.R.)	..	..	8 4	..	5 0	..
Newcastle O. T. (N.E.R.)	..	15 0	..	..	20 0	..
Cardiff (G.W.R.)	..	..	9 2	..	4 2	..

Rates from Wolverhampton include collection within prescribed limits.

Both rates to Liverpool include delivery alongside ship there.

The rates to Southampton are for shipment only.

Swansea to London rate includes barging in London within prescribed limits.

The Swansea rates do not include collection in Swansea.

ADVANTAGES OF SOUTH WALES.

The main question to be considered in drawing a comparison between the manufacturers in South Wales and manufacturers in the Midlands are as follows:—

1. Cost of materials.
2. Facilities for shipment.
3. Cost of labour.

Cost of Materials.—In my opinion, steel bars for rolling into sheets for galvanizing and corrugating stand *first;* and the facilities and means of production in South Wales are considerably greater than those of South Staffordshire. A confirmation of this is shown by the fact that several Staffordshire firms have closed their forges, having abandoned the manufacture of puddled bars, and are now buying steel bars from South Wales, paying carriage to their mills, galvanizing them, and afterwards paying carriage to Liverpool!

It is requisite to have a good surface to produce good galvanized sheets, the excellency and regularity of the coating depends on the perfection of the surface; so it must be apparent to anyone that a porous sheet of poor quality will absorb considerably more spelter than a sheet with a good surface, and steel bars produce an admirable surface for galvanizing purposes.

The general practice in South Staffordshire is to roll sheets from puddled bars faced with a bar of better quality, and the quality depends to a great extent on the prices ruling, as when the competition is severe the quality suffers accordingly.

Before my old firm rolled their own sheets we were purchasers, and amongst all the numerous sheet-iron makers in the Midlands I could not count on more than six who could supply a *regular uniform quality*.

As the consumption of steel bars for tin-plate making is more likely to decrease than increase, there is more probability that the demand will not exceed the supply, and that prices will not rush upwards with the spasmodic movements peculiar to the South Staffordshire district. This is shown by the course of business during the last twelve months, for, while sheets have increased in value in South Staffordshire, the same prices are quoted for steel bars in South Wales now as were ruling twelve months ago.

Spelter.—South Wales is practically well placed for this article, as there are several works there, while Staffordshire makers have to pay carriage from South Wales, Hull or London.

Labour in rolling would not exceed that of the Staffordshire makers, and the cost for under-hands should be less, considering the enormous number of men out of employment in South Wales. This would also apply to the labour in galvanizing.

Shipping Facilities.—Galvanized iron can be delivered in Liverpool by water from Swansea as low as 8*s.* per ton, while the railway rate from the Midlands, including dock and town dues, is 10*s.* per ton. I have no doubt that it can be transported by water and delivered f.o.b. London at a cost not exceeding 10*s.* per ton. Works in the vicinity of Swansea have the advantage of direct shipment from that port to many places abroad and the probability of the extension of these advantages eventually.

Home Trade.—There is a considerable business to be done with iron merchants and contractors, who pay monthly; and a remunerative price can be obtained from this class of trade. South Wales is well situated for this trade in Ireland, and West of England. There is also a good demand for iron roofs for collieries, iron-works and agricultural purposes; and this is a good outlet for galvanized corrugated sheets.

Manufacturers in South Wales, having works with the requisite power and the necessary strength of housings to receive rolls for rolling sheets from steel bars, stand in a first-class position for manufacturing galvanized corrugated sheets. The tin-house generally can be adapted for a galvanizing shop, and, in fact, there are no works so capable of being readily adapted for the manufacture as the existing tin-plate works.

HOW TO SUCCEED IN THE BUSINESS.

As these experiences have been written more especially for the perusal of the manufacturers of South Wales, I have treated the subject of the advantages of that country for carrying on this manufacture under a separate heading, but it should be read in conjunction with the following as one of the elements requisite to ensure success of the undertaking.

The relative position of manufacturers in the galvanized iron trade to-day is as follows:—

1. Firms producing sheets and galvanizing them on the same premises.
2. Firms having the sheet works and galvanizing works separate, and at a distance from each other.
3. Firms buying their sheets, and galvanizing only.

A Concentrated Works, rolling its own sheets and galvanizing them on the same premises, must evidently stand at a less rental and dead charges than that of separated works, and this is without taking into consideration the necessary haulage between the two works, which is equal to 2 or $2\frac{1}{2}$ per cent. on the selling price. *Therefore, one of the chief elements of success is to have the sheet rolling and galvanizing on the same premises.*

The Dead Charges should be rated at the lowest possible figure, but these are figures to be determined by the manufacturer himself; the one who, rolling his own sheets and galvanizing them on the same premises, rates his dead charges on a moderate scale, must necessarily stand in a unique position, and, having this advantage, it is such as will enable him to adopt the proper methods to ensure a quality that will secure a permanent and remunerative business, and make a profit against inferior makers.

The next Element of Success is to cultivate the business of those markets into which, by getting a "footing," he ensures regular indented orders for his brands, thereby regulating the production from his mills of sizes and gauges which he would find in time he could rely upon being called for regularly. I have treated the subject of the sizes and gauges required for the various leading markets of the world under another heading.

A Properly Classified List should be in the possession of the manufacturer, of the various buyers in London, Liverpool, Birmingham and elsewhere, showing the markets they ship to, as far as can be ascertained. There are firms abroad having their branch houses or agencies here, and there are firms who buy for all markets. Again, there are agents and brokers who buy for the foregoing buyers. These should all be carefully discriminated. When a specification gets into the hands of an agent or broker, the manufacturer is likely to receive the same enquiry from several sources, showing it is going the "round of the market." These are specifications which can only be secured by bottom prices, irrespective of quality.

The Best Plan is to have a list of the various markets, and the buyers' names under each market, for reference. For instance, the chief and largest buyers for Valparaiso (Chili) are Balfour, Williamson and Co., Liverpool, and Rose-Innes and Co., London, and I suppose the bulk of the galvanized iron landed there is in their names. They each hold more stock than many galvanized iron manufacturers hold.

Cabled Orders. — It often happens that merchants abroad keeping stocks of galvanized iron, having had a heavy demand on the stock, or on special sizes, are unable to wait the delay consequent upon sending the order home to the branch house or the agent here by the mail, and therefore cable the order, which is invariably required for immediate shipment. In such cases a buyer will often give an enhanced price to a firm upon whom he knows he can rely for prompt despatch.

Prices Current.—Some manufacturers make a practice of circulating a printed price list monthly, and quoting prices which they would be only too happy to accept! It is an understood thing that such prices contain a margin, so that it is of little use to circulate printed matter containing information which can readily be obtained from the trade journals every week. If it is done with the object of "keeping the name before the buyer," a well got up card, showing the brands, etc., will answer the same purpose, and at considerably less expense.

Having ascertained the requirements of buyers, as explained above, the next consideration is to ascertain the capacities of your mills as to sizes and gauges, so that you know for a certainty the class of trade it will pay you best to cultivate.

Competent Agents.—As this is really one of the most important features, and especially in a new undertaking, it is highly desirable that a man should be selected of whom it is quite evident he will have his employer's interest at heart.

It is also requisite that he should know that the employer is fully acquainted with the requirements of the various markets, and that he is not only fully conversant with the cost of manufacture, but that his works are in a position to do the trade, and do it well, while at the same time he does not intend to make a sacrifice of his advantages to the incompetency of others.

There is this difference between galvanized iron and tin plates, viz., that the lowest prices do not always secure the business. That they do secure a certain amount of business there is no doubt, but the class of business worth cultivating is that which will in time ensure you regular indented orders.

The Agent Must Live, either by salary or commission, and it is a question for the manufacturer to consider if it is not more advisable to pay a salary to a competent active agent, or part salary and part commission.

I remember being in a merchant's office when the agent of a galvanized iron manufacturer called, saying his works had instructed him to sell two or three hundred tons of galvanized iron, as they wanted orders badly. He named the price, which was considerably under the bottom prices. The merchant had no orders, and the only result obtained was a "slaughtering" of the prices for the next comer. This agent was on commission, and I suppose he was in want of some money. My experience is, that it does not require a vast amount of competency on the part of an agent to sell lower than everyone else. It is not a matter for surprise that the works represented by this agent ultimately went into liquidation.

There is always plenty of business to be

obtained at remunerative prices, but it must be obtained in a legitimate manner.

Fully Posted Up.—It is very essential for a galvanized iron manufacturer to be posted "up to date" in prices and current events of the trade, as contained in such journals as the 'Ironmonger,' the 'Iron Trade Journal' and 'Ryland's Iron Trade Circular'; and money is well spent in subscribing to several of the Australian prices current, as he can readily gauge the state of the markets, etc. He should also be conversant with the capabilities of his competitors, to enable him to judge of the force of any competition he may meet with from time to time.

Advertising.—A considerable amount of money is injudiciously spent by some manufacturers in this direction, and this is chiefly due to the want of experience on the manufacturer's part, or by his having been led away by the plausibility of some advertising agent. The number of English "trade" journals professing to have a circulation abroad is legion, but it must be inferred that if they are to be circulated they must be subscribed for, or they must be sent abroad for gratuituous circulation. I leave it to the reader to decide which is the most feasible!

The Best Way to Advertise is that of advertising in the chief journals circulating in the country whose trade you are desirous of cultivating. The charges are considerably lower than those of the English journals, and there is more certainty of a manufacturer's advertisement producing good results than if it appeared on this side. I would recommend the 'Melbourne Journal of Commerce' for Victoria, and the 'Sydney Trade Review' for New South Wales, and a similar class of publications for other places. Publications of this kind have this advantage, that they are regularly mailed home, so that the buyer on this side is certain to see the advertisement. The best plan is to have the advertisement "set up" here, and a stereo made of it and sent out, so that you are certain of its being satisfactory. An English buyer, as a rule, does not gain his knowledge of manufacturers from the English trade journals.

A Visit Abroad.—Where there are two or more partners in a business, there is no form of advertising that can produce such good results as by one of the partners taking a trip abroad to visit the leading markets. If a partner cannot be spared, a "live" accredited representative should be sent who has a good knowledge of the business. I venture to say that a trip of twelve to eighteen months, visiting the Cape, Australia, New Zealand, and returning by the west and east coasts of South America would produce in two years more profitable and permanent results than ten years of advertising. It is much easier to gain access to the principals abroad than it is on this side, and they are always pleased to interview visitors from the "Old Country." It also gives such facilities for seeing what your competitors are doing, and finding out your own capabilities. New ideas spring up, which must bear fruit on the return home. The late John Lysaght made three separate visits to the Colonies, and it was owing to this fact that he laid the foundation of the present magnificent business. I need hardly say that the facilities for travelling are now so perfect, and the comforts so complete, that ocean travelling is equal to living in a first-class hotel, with the advantage of reaping considerable benefit from the intercourse with experienced fellow-travellers.

Trade Combinations.—There have been several attempts made to obtain a combination of all the galvanized iron trade, with the double object of raising prices and also to keep new works from starting. I have no faith whatever in such combinations, because they generally emanate from people who take the first opportunity of breaking the restrictions to their own advantage. Some years ago there was such a combination attempted, but it only lasted a few months. My experience is, that it is not only an impossibility but it is not politic to be bound by the restrictions necessary to effect such a combination, and especially so as regards a manufacturer who may be just commencing in the trade. It will be found, as a rule, that these schemes are fostered by firms who, from various causes, as already explained, find they are losing ground, and are unable to compete with firms whose advantages are greater. This being the case, they exercise their ingenuity to bring about such a combination as will effectually bind the newer firms from availing themselves of their advantages, to the benefit of the projectors themselves.

A manufacturer whose works are arranged on a solid basis does not actually feel the necessity or advisability of having his hands tied down for the benefit of others who may not be so well situated as himself. I would say emphatically to such manufacturers, Do not entertain any such proposals should they arise later on. It is a *suicidal policy* for a new beginner to enter into any such arrangement.

Buyers do not like combinations, and human nature is the same now as it ever has been. I do not doubt there are manufacturers who would enter into such an arrangement with a view of abiding by it honourably; but I fear also that there are others who would obey it in the letter but not in the spirit. This forms an obstacle which alone is fatal to its success.

Consignments.—I shall treat later on respecting this ingenious method of doing business, which is offered with a view of "opening the market" for you by sacrificing a third or more of the value of the goods. It is a costly and tedious operation if the goods require to be sacrificed in order to show the buyers that you are in a position to make a quality equal to that of any of the best of your competitors!

If you have to make a sacrifice, it is far better to do it at this end, with a merchant of standing, and know the extent of your loss at once; but making such a sacrifice is entirely useless unless the buyer has sufficient confidence in your ability to supply a good quality regularly, give quick despatch, and quote a fair price at all times. This confidence can only be gained by personal interviews, or through a trustworthy and competent representative.

These consignment schemers are a species of "land shark," and woe be to the manufacturer who gets within their avaricious jaws!

The London or Liverpool Office should be represented as really a branch of the business, and not as an agency, and the stationery of the office should bear the manufacturer's name and the address, so that the same importance would be attached to the communication as if it came direct from the works.

Good Stationery.—A good impression is created of the status of a firm by the effective manner in which the stationery is printed. Thoroughly good paper, with a neatly engraved heading, conveys a good impression, and a first impression is often of significant importance. The difference in cost between good paper and bad, between good printing and inferior work, is so little, it is a good investment to have this outward badge of the firm as good as possible. When once the design of note-paper, etc., is chosen, it should be adhered to.

Economy in Manufacture.—Whilst every economy should be practised in the mode of manufacture, it is false economy to endeavour to build up a remunerative and permanent business by reducing the quality of the goods.

Good Materials.—I have explained how the quality of the spelter can be ascertained, and how desirable it is to buy the best that can be procured; and this should also apply to all the other materials required.

Good Workmen.—Good materials in the hands of inferior workmen cannot produce satisfactory results; and a manufacturer who is sufficiently alive as to the policy of purchasing good materials, will also take the precaution to have competent workmen to manipulate them.

Night-work. — Whilst speaking of competent workmen, I would specially refer to the competency of the man who has charge of the galvanizing baths at night. Upon this man devolves the duty of having the baths in such a condition that the work can be commenced immediately upon the arrival of the men in the morning. If the baths are "set," or cold, a considerable amount of time will be lost in waiting for them to get hot enough for work; and, on the other hand, if too hot, they must be cooled down again, which also means a loss of time.

Night "Shifts."—I have never found good results from night-work in galvanizing. This arises from the difficulty of seeing any faults by gaslight as readily as by daylight, and also from the galvanizing fumes which, being light, are kept down by the colder state of the atmosphere at night, especially so in the winter; so that the tendency is for the sheets to lose their bright appearance. The employees too, expect to be paid as much for five nights as would equal six days, they do not care for continuous night-work, and difficulties are always arising in changing the labour about. In addition to this, female labour cannot be employed at night, so the cost of production is considerably greater.

Fair Price for Labour.—The question of labour I have gone into under another heading, in dealing with the costs; but I may remark here, that to have competent workmen it is necessary to pay a fair price for the labour. There is no saving in the engagement of incompetent men, or youths, as one or two damaged sheets in the course of a week will amount to considerably more than the difference between the two classes of men, whether applied to packing, dipping or corrugating.

Orderly Works.—The galvanizing trade is one that produces a considerable amount of debris in the shape of old flux-boxes, worn out flux-buckets, burnt sheets, etc., etc.; and unless a systematic plan is adopted from the commencement, the probability is that the works will speedily become encumbered, and every available corner filled up with this lumber.

The best plan from the outset is to set apart an unused and "out-of-the-way" portion of the works for the storage of such articles, and also for the flux-casks, with the object of keeping the works in a thoroughly orderly condition. While the baths are being prepared for work, the females should brush up the shop, and have imperative instructions to do this daily. The cost of this amounts only to the value of the stiff brushes required, and this is a trifle compared with the results to be obtained. It must be apparent that waste in manufacture, and consequent incompetency, can be more readily seen if a works is kept in strict order than it would be otherwise. Besides this, a manufacturer is liable to receive "surprise" visits from such buyers who may be anxious to ascertain his capabilities to make not only a first-class quality, but to do so *regularly.* To find such a works in a state of disorder must naturally create a very bad impression upon a visitor, and is also apt to create a doubt as to the standard of manufacturing, which he has been led to believe was the foremost quality of such a manufacturer's production. "A place for everything, and everything in its place," is an old saying, and a well-kept works is a pleasure both to the employer and his employees also.

Another point is, that the stock of muriate of ammonia and spelter should be kept in a store, which should be rigidly kept locked, and the supplies should be weighed out and booked daily. There should be a general store also, locked up, for other articles, and both of these could be controlled by the time- or gate-keeper.

To ensure economy in the consumption of muriate of ammonia, it is a practice with some manufacturers to give a small premium or bonus to the dipper on any ammonia returned to the store at the conclusion of the day's work, which has been saved from the amount allotted to him, *pro rata* for the quantity of the sheets he has had to dip. Providing that the quality has not suffered in consequence of such returns, I consider this system has a good effect, and is certainly an incentive to the dipper to use as little as he possibly can, consistent with the production of a first-class quality.

An account of such returns should be ascertained daily, and a general account weekly or fortnightly.

Wasters.—These should be assorted and cut down as they accumulate, those worth corrugating should be corrugated, made sound, and then stacked in heaps of the various lengths or gauges, and a record kept of the number of sheets in stock. Those unfit for corrugating should be cut up into straps and corner-plates for the cases, also for "trumpets" for the edges, and pieces for the ends of the bundles of corrugated iron.

Stock-Book. — Either a stock-book should be kept, so that the amount of materials and finished goods in stock can be ascertained at a glance, or the stock should be arranged in such a methodical manner that it can be taken very quickly.

Attention to Count.—It is highly important that strict attention is paid to the count of sheets required according to the standard of weights in the galvanized iron trade, and this, of course, depends on rolling them to the proper weight per sheet.

Careful Packing.—It is, as already explained, extremely important to pay strict attention to the packing, and also to see that no wasters are suffered to mix with the perfect sheets.

Patent Processes.—I shall describe the various alleged economical "patent processes" later on, so I will briefly remark that if you wish to make a success of your business you will shun these just as you should shun the consigning system.

A Competent Manager, who has his employer's interest at heart, is an absolute necessity. I shall show what can be expected when an incompetent man is employed. A competent man should know every detail, from the rolling to the packing, and should be a conscientious disciplinarian.

Cost Prices.—A scale of costs should be made, based on the rise and fall of materials, for ready reference, and it should be so comprehensive as to include every item of dead and other charges, so that the actual profit can be seen at a glance, and the "danger line" clearly shown when competition necessitates a reduction in prices.

Cash Payments.—I am a strong opponent of the system of either giving or taking bills in payment. The manufacturer who is so well placed as to be able to pay all cash, and in such a position as to decline to sell except for cash, is in a happier and safer position than the one who prefers the credit system.

HOW TO FAIL IN THE BUSINESS!

Some twelve or fifteen years ago there was such an ample margin of profit on galvanized iron manufacturing that second-class firms and new beginners in the trade felt themselves secure if they "cut" under the prices of the older and larger firms, whose prices they took as a guide, doubtless lacking the knowledge necessary for ascertaining the cost themselves.

Also, in those times, the number of manufacturers who bought their sheets and galvanized them was greater than the manufacturers who rolled their sheets and galvanized them also. In previous years limited companies have rushed into this trade who, without any previous knowledge, and without having ascertained the profits that have been realised by established firms, have hastily come to the conclusion that they had only to lay down the machinery and cut under existing quotations to realise handsome dividends. They have soon discovered their mistake, and have as hastily retired, perhaps sadder, but wiser men.

I only know of two limited companies that, having been formed for the purpose of manufacturing galvanized iron, have survived These are the Blackwall Galvanized Iron Co., Limited, London, and Alfred Baldwin and Co., Limited, Newport. In the former case this company have turned their attention principally to odd-work galvanizing, which, in London, is a profitable business. In the latter case, this is nominally a private firm. In addition to this Mr. Alfred Baldwin is a gentleman of wide practical experience, both in sheet-making and galvanizing; and of the latter he has obtained an extensive knowledge through being a director of the Blackwall Co., Limited, from its commencement. The speculative nature of the business, as referred to in my opening remarks, has now completely changed, and the galvanized iron manufacturer who does not know the cost of his productions had far better close his works and invest his money in Consols.

I venture to say there is scarcely any trade in which money can be more quickly lost than in the manufacture of galvanized iron, such results being due to several causes; and it is a significant fact that the following firms have failed, or gone into liquidation (some being reconstructed), during the last five years,

AND YET THE VOLUME OF THE TRADE HAS BEEN INCREASING YEARLY!

(1) Gospel Oak Iron and Galvanized Iron Co., Limited. Liquidation and reconstructed.

(2) Morewood and Co., Limited. Liquidation and reconstructed.

(3) Thomas Jordan and Co. Failed, and plant sold.

(4) Bruce and Still. Liquidation; business sold to a limited company.

(5) Wednesfield Galvanized Iron Co., Limited. Liquidation, and plant sold.

(6) Imperial Galvanized Iron Co., Limited. Liquidation, and plant sold.

(7) Morewood and Heathfield, Ltd. Liquidation, and plant sold.

(8) Thomas Bishton. Failed recently; result unknown.

(9) Thomas Skidmore and Co. Business closed, and plant sold.

(10) Avondale Galvanized Iron Co., Limited. Liquidation, and plant sold.

The liquidation of the first two firms, although the oldest in the trade, was mainly caused through their excessive dead charges, management expenses, etc.

The third, fifth and sixth firms and companies failed chiefly through their ignorance of the trade and cutting prices.

The third firm failed after an experience of about *eighteen months*, and the tenth after an experience of about *twelve* months!

The Globe Galvanized Iron Co., Limited, Great Bridge, and the Staffordshire Galvanized Iron Co., Limited, Walsall, are in addition to the foregoing list, and these concerns were started in what I may term the "golden age" of the galvanised iron trade. Both these concerns went into liquidation, and were sold to private firms, who have since carried them on successfully.

This leads me to investigate the causes of failure, and such an investigation is even more necessary now that the trade of galvanizing, to be made profitable, can only be carried on safely by taking a lesson from the failures of others.

The profits now are reduced to a remunerative profit, which can be derived from the rolling and galvanizing of the sheets being combined as one business on the same premises; but even these advantages can be counteracted from the following causes:—

(1) Badly arranged works, causing loss in manufacturing.

(2) Waste caused by incompetent management.

(3) Manufacturing without knowledge of the requirements of the markets.

(4) Excessive dead charges and management expenses.

Badly arranged Works.—If new works are to be laid down, they should be arranged on the most favourable lines, for ensuring economical consumption of fuel and saving labour. If existing works are to be rearranged and adapted for the manufacture, advantage should be taken to carry out the same methods.

ARRANGEMENT OF A WORKS.

"Many a mickle makes a muckle," is an old Scotch saying, and this is never exemplified in a greater degree than in the arrangement of a works.

I have inspected many works, and in all my experience I have never yet seen a well-organised works, laid down on economical lines, which has come to a stand; and it is to be inferred that a well organised works must necessarily possess a good manager. I visited a works a short time since, which had lost heavily the last few years, and had come to a stand. Of a group of four boilers, there were three good boilers (new), gauged up to 200 lbs. pressure, and the third was a smaller, second-hand boiler (I presume bought at a low price), gauged up to 150 lbs., and I suppose 100 lbs. would be the safe line for working.

All four boilers were coupled together!

The boiler flues were minus the usual cross-tubes, which effect so great an economy in the fuel.

The engines were in all parts of the works, one being quite 300 yards from the boilers, and this was the largest engine amongst them; another engine was 250 yards away, and another 200 yards, *and all the steampipes were uncovered!*

Four expensive Cameron pumps had been laid down, one for each boiler, whereas one would have sufficed with an efficient injector on each boiler.

The old type of annealing furnaces were in use, using coal, instead of slack or very small coal.

No waste heat from the annealing or heating furnaces was utilised for steam raising purposes.

The means of transport from one part of the

works to another necessitated, in many cases, a second handling of the production, and many other points too numerous to mention.

And yet the buildings, and roofs, had been erected very substantially and regardless of expense. Is it to be wondered at, that these small outlets should eventually lead to a loss in working?

It is a great mistake to sink capital in massive brickwork, or stonework, and heavy timbered, tiled, or slated roofs, and at the same time neglect the vital portion of manufacturing, viz. the economical working efficiency of the plant. Light iron principals, resting on cast iron, or H wrought iron columns, covered with corrugated sheeting, answer the same purpose for an ironworks, as the massive erections I have described, and at about a third of the cost.

The tendency of modern times is to have the works laid down on such lines, that you are at least equal to the best of your competitors, or the result cannot be otherwise than failure.

Waste caused by incompetent management.—An incompetent manager is dear at any price. The labourer is worthy of his hire, and a competent manager of a galvanizing works cannot be valued too highly. As I am now dealing with incompetency, I leave the reverse side for consideration under another heading.

Incompetency in management has its effect in the works, and in the office also.

GALVANIZED IRON.

In the works, it is conspicuous by an ill-advised selection of bad workmen, whose principal study is to turn out as much as possible in the least time possible, without any regard whatever to quality.

It is conspicuous in the careless setting, and consequent bad firing of the baths, leading to overheating of the metal, thereby causing dross and oxide, on the one hand, or by insufficient heating, ensuring a loss of metal by coating the sheets too heavily.

It is also conspicuous by waste of materials; by loss in taking off the flux and oxide from the baths; by carelessness in the packing, resulting in damage, and consequent claims for damage on the arrival of the goods at their destination.

It is conspicuous by the accumulation in heaps of sheets of all sizes and gauges, wasters, etc., of which no record has been kept; and also by the accumulation of lumber, old iron etc., in every available corner of the works.

It is a difficult matter to say where incompetency begins, and where it ends, but it is an old axiom, that "like master, like man," and it is very certain that an incompetent manager will very soon have around him none but incompetent and careless workmen.

Manufacturing without knowledge of the requirements of the markets.—This applies to count of sheets per ton, the quality required, the corrugating, and, in fact, all the many technical points, necessary for a successful galvanized iron manufacturer.

Excessive dead charges and management expenses. These have been chiefly through purchases of land, building, and machinery, in times when these points were lost sight of in anticipation of the prospective profits to be realised. This also applies to charges for management, London office expenses, etc. Such changes have taken place in the galvanized iron trade during the last ten years as could not have been foreseen by its older members. I have known as much to be given for land, buildings, and galvanizing plant only twenty years ago (exclusive of working capital) as would now run two mills and a galvanizing plant on the same premises. It is an impossibility for a profit to be made now on the ideas current at that time; and the new beginner takes a lesson, and the older manufacturer has to lose his capital, sunk "unwisely and too well."

Consignments.—There are firms in London, and firms abroad having agencies in London, who place the bait before galvanized iron manufacturers, of "getting into the market" by sending the iron on consignment. I never knew a consignment account yet but it ultimately resulted in a loss in the end. The policy is, to show a profit on the first few consignments, with the object of drawing the unsuspecting manufacturer deeper into the net, on the "spider-and-the-fly" system. When the manufacturer finds he has reached the "end of his tether," the remaining consignments are certain to show no profit, and perhaps he has only received the usual two-thirds advance. I have known many cases, where the manufacturer has had a claim for a balance against him, so that he has not received two-thirds of the value of his goods, after waiting for nine, or twelve months. "The game is not worth the candle," because the "market" has not been opened after all: you have only opened your pocket, for others to take out part value of your goods, without their risking a penny. And this system of trading has brought several manufacturers to grief.

Another feature of the consigning business is, that of "joint accounts" in consigning. An offer is made, that if the manufacturer will consign a certain quantity, the proposer will buy an equal quantity. This is a modified form, and is on the principle, "heads I win, and tails you lose!" The manufacturer can rest assured, that if the market is favourable on the arrival of the iron, his assumed friend will purchase it at his own figure—of course, charging the commission, interest, etc., etc. If the market is "*down*," anyone can have it, at any price. As a rule, it is offered indiscriminately, "price being no object, to make room for other goods."

Incompetent Agents.—I have known manufacturers who have placed themselves unreservedly in the hands of their agents, who, having the sale of the iron on commission, have only one object in view, and that is to take the orders at any price, giving as a reason that another manufacturer is quoting the same price, etc. This indiscriminate "cutting of prices" has brought several manufacturers to grief, they having chosen to believe all the statements of their agent, which have ultimately proved to their own prejudice and final collapse.

Patented Processes. — I shall describe three patented processes in another part, so it is only necessary to say that my experience is, that a patent has generally been adopted with the object of producing cheaper than other manufacturers, quality being a secondary object with such manufacturers. I do not know of any firm who have been successful in the adoption of such methods. It is a suspicious matter when a manufacturer brings out a patent system, proclaiming that it will produce a fine quality and at a less price than any other system. If such pretensions were based on *facts*, it is evident that it would pay that manufacturer to keep his system for his own working, instead of letting his competitors have the "privilege" of working it on royalty.

REQUIREMENTS OF THE VARIOUS MARKETS.

The Australian and New Zealand requirements in galvanized corrugated sheets are 5, 6, 7, 8, 9 and 10 feet lengths by 8-3 inch corrugations, and 27 gauge (nominally 26 gauge). A small proportion of 24 gauge is sometimes ordered for curving into water tanks and roofs. It is highly important that the quality have a uniformly bright and spangled appearance, and the corrugations should be such as to match the leading brands sent there. As 27 gauge has a tendency to spring out of the dies in the first operation, it is essential to pass them through a second machine, the dies of which should be always planed to an iron template of the proper depth of corrugation, and, when once the proper size is ascertained, it should be adhered to. If these points are not maintained, it will be impossible for any new manufacturer to obtain any permanent footing in these markets. Each sheet should be carefully marked with the manufacturer's brand, which should not be of too large a size, as the marking should be quickly performed. This can either be done by stencilling or with an india-rubber stamp.

The requirements for this market in flat galvanized sheets are chiefly for a quality suitable for bending into spouting and ridge capping, as these goods are made up there. The sizes are 72 inches long by 24, 30 and 36 inches wide, by 24, 26 and 28 gauge. This iron should be as flat as possible and nicely spangled in appearance. There is also a demand for a superior quality of iron for working up into every purpose, and the sizes run 72 inches long by 24, 30 and 36 inches wide, by 18, 20, 22 and 24 gauge.

South Africa.—This market takes principally galvanized corrugated sheets in 5, 6, 7, 8, 9 and 10 feet lengths by 8-3 inch corrugations and 24 gauge. The quality should be uniform and good, and with a nice spangled appearance. There is also a good demand for ridging, gutters, pipes, screws, nails and washers.

South America.—The demand for Chili and the Argentine is about the same in size and gauge, but the mode of packing varies (*see* paragraph on Packing). For Chili the bulk required is 6, 7 and 8 feet lengths by 8-3 inch corrugations and 24 gauge, while for the Argentine a demand more or less exists for a proportion of 9 and 10 feet lengths in addition. There is also a demand from Chili for galvanized sheets corrugated with a small corrugation and 26 gauge, and these are used for the fronts of houses there. Gutters are also required, and as the rainfall is heavy, these run in sizes from 6 to 9 inches. There is at times a demand for black corrugated (ungalvanized) sheets, also for galvanized ridging, screws, nails and washers. A minor quantity of flat galvanized sheets is required, and these are generally of a special quality.

Brazil.—This market requires 6, 7 and 8 feet lengths, by 8-3 inch corrugations, by 28 and 30 gauge. A specified count of sheets per ton is generally given, but not so much importance is attached to the quality.

Manila.—The requirements for this market are generally 6, 7 and 8 feet lengths by 10-3 inch corrugations, and 24 gauge. This market is a difficult one to get a "footing" in, as the exports are in the hands of a few merchants here, and a manufacturer must show fairly good proof of his ability to supply a good uniform quality to suit the requirements of the buyers.

Italy.—The demand from this country runs chiefly on flat galvanized sheets, 72 inches long by 1 metre ($39\frac{3}{8}$ inches) wide, by 18 to 28 gauge. They are required as flat as possible, but of the ordinary corrugating quality.

India.—The requirements for this market are principally for 6, 7 and 8 feet lengths by 8-3 inch, 9-3 inch and 10-3 inch corrugations, by 18, 20, 22 and 24 gauge. The quality is not an imperative consideration, beyond being well coated, but the competition in prices is keen. There are also various contracts advertised from time to time, inviting tenders from manufacturers for the requirements of the various Indian railways. The mode of packing for these contracts is generally shown by a drawing accompanying the specification. It is generally of a cumbersome and expensive design, being based on the "red tape" views of the officials, and very often forms an important item of the cost. The sheets are always inspected before packing, and as the sizes are generally *special* ones and the manufacturer is entirely at the mercy of the inspector, it requires a considerable amount of circumspection in dealing with these contracts. I need hardly say the quality should be such that the inspector has no ground whatever for finding fault with the goods, or any pretext for asserting his authority. There is a considerable amount of difference between inspectors, and some will find fault merely to show they are doing "something" for their salary.

Canada.—This market only buys flat galvanized sheets, which are chiefly used for roofing purposes. The sizes are 72 inches long by 24, 30 and 36 inches wide, by 24, 26 and 28 gauge, the demand running chiefly on the latter gauge. It is highly essential that the sheets be cold rolled and close annealed, and they must be *flat*, being required for laying on boards for roofing purposes. The sheets must be *soft* and malleable, to stand folding, so that the joints can be "locked" in laying them on the roof.

Foreign Railway Contracts. — There are also contracts at times for the requirements of the Mexican, Chilian and Argentine Railways. As a rule, the inspection is not quite so rigid as with the Indian State Railways. It can be assumed these contracts are always for sheets of 18 or 20 gauge.

Home Trade. — The requirements for consumption in Great Britain are for all sizes, but chiefly of 24 gauge. The home trade will also be found a great outlet for " wasters," which, if made sound with solder, will, in 2-ton lots and upwards, realise the export price of the perfect sheets. In small lots a better price can be realised, and the best policy in this case is to sell them at per sheet, instead of by weight. There are special sizes required by the railways, chiefly in the strong gauges, also by contractors for Government work.

A large and regular demand exists for flat galvanized sheets of Best Best Quality, 72 inches by 24, 30 and 36 inches wide, by 18 to 26 gauge, for working up. This iron is chiefly bought by the wholesale iron merchants, who distribute it in small quantities. A very good reliable quality is required, and a compensating price can always be obtained.

THE BEST MARKETS TO CULTIVATE.

The Primary Object should be to cultivate the business of those markets whose requirements ensure the production of sizes and gauges which not only suit your mills, but also ensure a regular production, thereby enabling you to manufacture with a certainty of the sizes being called for regularly. Coupled with this, the markets should be cultivated which appreciate a good quality, and for which "top prices" can be obtained.

The markets most desirable to cultivate are the most difficult to enter—where quality is a great consideration; but the after results are that you will find your brand indented for with other leading brands; and this fact often reduces the competition to perhaps two or three makers.

The Cape, Australian and New Zealand Markets are the best in these respects as regards galvanized corrugated and flat sheets, and the Cape market as regards galvanized corrugated sheets. As already stated, the Cape market takes principally 24 w.g., 5 to 10 feet long, by 8-3 inch corrugations, 24 w.g., and the Australian and New Zealand markets principally 5 to 10 feet by 8-3 inch corrugations, 26 w.g.

The South American Market takes principally 24 w.g., 6 to 10 feet long, 8-3 inch corrugations, but brands have not such a hold on this market as with the English colonies. At the same time a good quality *is appreciated*, and if the manufacturer can gain the confidence of the buyer by showing proof of his ability to supply a good quality regularly, he will often have the preference, at an enhanced price, over makers upon whom such reliance cannot be placed.

India, although a keen buyer, is a good outlet for singles, and as the manufacturer must roll some singles at the commencement of the week, the requisite sizes will find a desirable outlet in the direction of this market, and can be stocked with this object in view.

Manila is a very good market to cultivate, and good prices can be obtained, as quality is really the *primary* consideration with the buyers. A maker of an inferior, or second-class quality, stands no chance whatever, simply because the lowest prices will not obtain the orders.

Canada is a good market for flat galvanized sheets, and good prices can be obtained when the brand gets known. A good plan is to send out half a dozen sheets of assorted sizes in a close felt-lined case to each of the leading merchants in Toronto, Quebec and Montreal for them to see the quality. No advertising is desirable beyond this, and if they approve of the quality they will indent for your brand through their English house immediately. There are only a few English galvanizers who can supply sheets good enough for this market, and the chief essential is the flatness.

METHODS OF PACKING, AND PRICES.

The mode of packing varies according to the market, and the following is a description for the leading markets, and the present prices obtainable over and above the price of the galvanized sheets :—

Market.	Mode of Packing.	Present Prices.
Australia and New Zealand	Half-ton of close felt-lined cases	15*s*. per ton.
South Africa	Bundles of about 2 cwt. with felted edges	No charge.
Chili	Bundles of about 2 cwt.	No charge.
Argentine	Skeleton cases about 5 cwt.	5*s*. per ton.
Monte Video	Skeleton cases about 5 cwt.	5*s*. per ton.
Brazil	Skeleton cases	5*s*. per ton.
Manila	Half-ton felt-lined close cases	15*s*. per ton.
India	Bundles about 2 cwt., also	No charge.
	Skeleton cases about 10 cwt., also	5*s*. per ton.
	Skeleton cases about 5 cwt.	5*s*. per ton.
	Also cases made to specification.	
Italy	Bundles about 1 cwt.	No charge.
Canada	Close cases about 5 cwt., also	15*s*. per ton.
	Close cases about 10 cwt.	10*s*. per ton.

Cases for curved sheets are extra, according to curve.

For the South African Market it is the usual practice to pack galvanized corrugated sheets in bundles, with felted edges, to protect them from damage in transit. Strips of galvanized iron are cut 4 to 5 inches wide, and a strip of felt the same width. The iron is then bent to clip the edges of each side of the bundle, and the strip of felt inserted inside it. The cross bands go over these strips and, when closed down, hold them fast. For the ends strips are required 9 to 10 inches wide, and a piece of felt the same width. The iron is bent to clip the end and the felt inserted, and the two longitudinal bands make them secure.

This is a favourable opportunity for using up "wasters," as all gauges can be used for this purpose, and it also disposes of odd pieces of felt. It is usual for the hoop-iron banding to be $1\frac{1}{4}$ inch by 12 gauge. The quality should be very good, to allow of the ends being hammered down close. It is usual to put three bands on 6, 7 and 8 feet bundles, four bands on 9 and 10 feet bundles. Some manufacturers do not galvanize the bands, but as the oxidation of the bands is likely to stain the sheets, I do not think it advisable policy to pack the bundles with black ungalvanised bands.

Skeleton Cases are generally made with $1\frac{1}{4}$ inch sides and 1 inch ends, and battened with 6 or 7 inch by $\frac{3}{4}$ inch battens, each of these being strapped. The sizes are easily ascertained by taking the quantity of sheets required to be packed, and allowing 1 inch at the sides and the same at the ends.

Felt-lined Cases are generally made with $1\frac{1}{2}$ inch sides and $1\frac{1}{4}$ inch ends, covered with $\frac{3}{8}$ inch boards, which should be grooved for wood tongues. Allowance should be made for the felt lining, which, with an allowance of space for packing, should be $1\frac{1}{4}$ inches over and above the width and length of the sheets, and $\frac{1}{2}$ inch over the depth of the sheets. The felt used is sarking felt, and as this varies in substance, etc., it should be selected of a pliable quality, and while it is required thin it should not be so thin as to show the daylight through. It is the best policy to have a sample roll for inspection before purchasing a quantity. I found by experience that McNeill and Co., Bunhill Row, London, made the most reliable quality for the purpose. The felt is cut of the required dimensions of the case, with an allowance for turning up two or three inches at the sides and ends, the corners being notched for the purpose. First, coat the interior of the case with a mixture of hot tar and pitch, and, having warmed the felt, it will readily adapt itself to the shape of the case. A few flat-headed wire nails are required to nail the edges of the felt.

For the purpose of warming the felt a stove should be built with a cast-iron plate on the top 6 feet by 3 feet, by 1 inch thick, and a very little coal will suffice to keep it hot.

When the case is filled, a piece of felt the outside dimensions of the case, is placed on the top, and the boards nailed through it. It is requisite to line the case with cheap brown paper to prevent any tar from coming into contact with the sheets.

I have refrained from giving the sizes of the cases, as this all depends on the depth of the manufacturer's corrugating discs, and also on the competency of the corrugator, and I would advise a manufacturer to test the sizes of his cases by making one of each before making a quantity. When once the sizes are decided upon, there need be no alteration.

It is not usual to put strips (or felt) on the edges of bundles for the Chilian market, but, with a view to utilising the "wasters," it is not a costly experiment for a manufacturer to put on the longitudinal strips (without the felt) even if they should be objected to, and the weight deducted afterwards. It could be tried as an experiment. It acts as such a great preservative to the edges of the bundles, and makes them so much easier to handle, I feel sure very few, if any, buyers would object to the slight weight of them.

Gutters and Ridging are generally packed up in skeleton cases, containing 100 lengths each, and the cost of the case is included in the price quoted for the goods.

Pipes are generally packed in cases of 50

lengths, according to the size. If several sizes are ordered they can be packed 100 or 150 lengths in a case, as one size will go into the other. These cases are also generally included in the price.

Importance of the Packing. — I cannot too strongly impress upon a manufacturer entering this trade the importance of strict attention to the packing. It is useless to manufacture a good quality unless he can ensure his goods arriving at their destination in good merchantable condition.

From personal observation of the landing of galvanized iron at Buenos Ayres and Valparaiso (Chili), I have felt the importance of pressing this point on the attention of galvanized iron manufacturers. I have seen the bundles landed with half or three-parts of the bands off, and the cases absolutely falling to pieces on landing, through the inferior nature of the cases.

It is highly essential that the hoop-bands should be of such a quality as to ensure freedom from breakage, and the cases should be made of sound, well-seasoned wood.

When the output of a galvanizing works will warrant it, I consider it advisable to put down a saw-bench, and keep an assortment of spruce deals in stock, and so ensure a supply of well-seasoned timber.

BRANDS.

I would advise any manufacturer who may be commencing in this trade not to have a multiplicity of brands, as it is a great mistake. In choosing a brand great care is required, and the first consideration should be to make a thorough search of the Trade Mark journals, in order to ascertain all the brands that are registered by other manufacturers. In addition to the brands of manufacturers there are many private brands used by merchants and agents. Having ascertained this, the next step is, not only to choose some design that shall be original and effective, but also capable of being easily and effectively stencilled on the sheets. It is not desirable to use a crown, or the words "crown brand," either singly or in combination with any other design or letters, as it is being used by so many manufacturers, and, being a mark common to the iron trade, it cannot be registered; besides this, it is *not* now a symbol of a good quality by any means. The design having been fixed upon, the next step is to register it, and as this takes about three months before it is clear of the Trade Mark journals, it is therefore desirable to decide on the matter at the onset, or a considerable delay may ensue before the manufacturer can commence to use it with safety; and for this reason it is essential to select, as far as possible, a design that he considers should pass the registration.

I would suggest one brand for corrugated and plain sheets of the same quality, another brand for the working-up sheets, adding for the third quality, the word "charcoal."

The branding is generally done by stencilling, or india-rubber stamp, but the latter is liable to get injured from the heat of the sheets. I consider stencilling the quickest mode, and if the brand is cut of a moderately small size it can be easily and quickly done, and it should be done while the sheets are warm, so that the ink will be quite dry. I am willing to assist any one in the selection of a brand, and if a search is required I would merely charge for the time taken up in the search.

The Use of Private Brands for some buyers is a question that must be left to the manufacturer's discretion, but in my opinion it is not the class of trade to be encouraged by a maker of a thoroughly good quality, unless he has a guarantee that he alone will have all the business that may accrue from it. If he has not this guarantee he is simply creating a trade for the owner of the brand, who will leave him when it suits his purpose. When a private brand is proposed, and with a view to secure the business, it should be registered conjointly by the manufacturer and the owner. However, even in its best light, I consider it is a practice that galvanized iron manufacturers should not encourage by any means.

CORRUGATING OF GALVANIZED IRON.

The corrugating of galvanized sheet iron is effected by means of a powerful machine, in which are placed two dies which fit into each other, the bottom die being wider on each side by half a corrugation. The most useful size for a corrugating machine is 10 feet 6 inches between the standards, to take dies 10 feet 3 inches long for corrugating 10 foot sheets. From my experience I am not in favour of a machine of greater length as the bed is apt to "spring," and although machines are in operation for sheets up to 12 feet long, my impression is that the limit of a machine should be 10 feet 6 inches in length. The demand for 11 and 12 feet sheets is so small in comparison with 6, 7 and 8 foot lengths, I do not consider a manufacturer would be prejudiced by professing his inability to corrugate over 10 feet in length. The "hammer" of the machine, which contains the upper corrugating die, is operated by means of eccentrics or a cranked shaft. The former is decidedly the best plan and less liable to breakage. I have known a crank to break through, and the shaft is then useless as it cannot be repaired, and would take some time to replace. In case the straps of the eccentric break they can easily be replaced. The sizes of dies used are the 3-inch dies for bulk of the work performed. Also 1-inch dies for corrugating sheets for the Chilian market. These sheets are used for the fronts of houses, and are more effective for this purpose than the usual size. The other special sizes of dies are 2, $3\frac{1}{2}$, 4, 5, 6 and $9\frac{1}{2}$ inch. In addition to these a "stepped" corrugation is made for corrugating, or rather indenting, sheets for filling in the sides of bridges, etc. There is not a very great demand for this kind. It is worth the attention of a manufacturer to produce a corrugation or indentation in imitation of weather boarding for use on the sides and ends of iron buildings. This would be more effective than the usual 3 inch corrugations, and if painted a stone colour would relieve the monotony of iron structures. In addition to the above there is the "Italian" form of corrugation, which is merely a plain roll on each side of the sheet. Messrs. Braby and Co., Limited, also produce a special form of corrugated sheet which is worth remark. The outside corrugations are slightly larger than the others, and the patentees claim that it is proof against rain blowing through the overlapping joints.

The machine is operated by two men, the corrugator and his underhand, who as a rule do the work at piece-work prices. It appears a simple operation to corrugate a sheet of iron, and yet it is not so easy as it appears, and an efficient corrugator only attains his skill after a considerable amount of practice. The chief requisite is confidence, and the alertness necessary to prevent the fingers from being corrugated in addition to the sheet iron. The sheets are corrugated in pairs in 22 and 24 w.g., and three together in 26 and 28 w.g., while 20 w.g. and thicker are corrugated singly. The chief aim of the corrugator is to press the sheet forward to the side of the die, and of his assistant to aid his effort and pull it forward in the same direction. If the sheet was merely worked through the machine without a fixed method the result would be that the corrugations would vary, not only in width but in depth also, and the sheet iron would consequently be unsaleable. The corrugating can also be done by means of a machine with corrugated rollers, but this system is rarely used. I do not consider it so effective as the usual form of corrugating machine.

Some few years ago there was a considerable discussion in the Australian trade journals as to the variation in the sizes of corrugations of the different brands of galvanized corrugated iron. The following leading article on the subject appeared in the *Australasian Ironmonger*:—

THE CORRUGATION OF GALVANIZED IRON.

An interesting letter on this subject appears in the *Ironmonger* of Nov. 27, written by Mr. James Davies of London, formerly managing partner for fifteen years of one of the busiest galvanizing works in the Midlands. Mr. Davies therefore gives us a manufacturer's view of the question, and winds up his argument with the assertion that though to have uniform corrugations appears to be a simple matter at first sight, yet it is beset with innumerable difficulties *unless the proper course is followed out.* [The italics are ours.] And we thank Mr. Davies for his letter, which, written in quite the opposite strain, turns out to be one of the most powerful arguments in support of the movement that has yet reached us.

According to Mr. Davies' own showing, uniformity *to a certain extent* is almost indispensable if galvanized corrugated iron is to be sold in the Australian markets. He speaks of the difficulties he had to contend with until he was able to make the corrugations of his brand follow the leading brands in the Australian market. This is all any one could ask for. Decide on some standard—that of leading brands, if you like—and then adhere to it. How completely this answers the objection raised by some manufacturers, that, by adopting a standard gauge, small makers would be placed on the same footing as large ones. If there is any advantage to the manufacturer in this, cannot the small maker at once adapt his dies to the most popular gauge?

The difficulties Mr. Davies describes may be reduced to two. First, the dies for stamping the corrugations get worn, so that the tendency is for the gauge to widen; secondly, it is difficult to secure uniformity of temper and softness. The first of these difficulties is an old objection to uniformity in everything. Wire gauges, screws and gaspipe gauges all get worn, and constant watchfulness is required to replace them on the first appearances of inaccuracy. Mr. Davies tells us, indeed, how easily he met this difficulty. He kept a steel template of the corrugation to which the dies were planed on the premises, and those in use were replaced by new ones as soon as they showed signs of wear. How simple it would be to plane the dies to a standard corrugation!

The second difficulty is as follows. " Unless the sheet iron is uniformly softened it is impossible to corrugate the sheets uniformly, and, however deep the dies are, the sheets spring back out of the dies. It is useless arranging for corrugations of a certain width and depth unless the sheet will stop in the corrugation when forced into the die in the process of corrugation." This is self-evident. But Mr. Davies, after condemning two processes, describes a third, which gives the necessary uniformity, namely, close annealing in large wrought iron boxes. It is true that this process is expensive, and will not pay when the output is less than fifty tons a week; but it is practicable. From Mr. Davies' own showing, therefore, nothing but want of manufacturing skill prevents the maintenance of a uniform gauge, when once that gauge is decided upon. The expense of altering the dies must be comparatively trifling. They must need continual replacement, and, therefore, a standard could easily be adopted at one of the periodical renewals. The maintenance of uniformity in corrugation is, as Mr. Davies says, "A question of ability to do it," that is to say, of manufacturing skill. The arguments which were used at first against the adoption of a uniform gauge for the corrugation of galvanized iron are gradually melting away. Mr. Davies' contribution to the discussion is valuable, for it shows that, from a manufacturer's point of view, all that stands in the way is obstinacy or carelessness in neglecting to adopt a uniform gauge for the dies, and want of skill in securing uniformity in the softening of the sheets. Messrs. M'Pherson, in their last letter to us, attributed the want of uniformity to the stiff-necked policy of the manufacturers. We extend the charge as above. As Mr. Davies says, it is no question of monopoly; and if the consumers, who are the real arbiters in the case, are sufficiently interested to make their demands felt, uniformity of corrugation will soon be an established fact.

I have already referred to this under the heading of "Sizes Required for the Various Markets," but this testimony as to the importance of the question is *specially valuable*, coming as it does from the market itself. To any one who has not any practical acquaintance with the trade, it would appear that one corrugated sheet is just the same as another, and that it is impossible for there to be any difficulty in the fixing together of different brands of galvanized corrugated iron. *A difference of $\frac{1}{16}$th of an inch in the width of one corrugation will make a difference of half an inch in the sheet, and 50 sheets side by side would show a difference of 25 inches.* Any manufacturer, therefore, who intends to have part of the Australian trade must essentially pay specific attention to the corrugating, so that his sheets will fit, without any doubt whatever, the corrugations of the "Orb" brand, manufactured by John Lysaght, Ltd., or the "Mars" brand, manufactured by George Adams and Sons, Ltd., or the "G O" brand, manufactured by the Gospel Oak Galvanized Iron Co. These are now the chief brands, and the corrugations of these sheets will be found to be alike.

ROLLING FROM STEEL BARS.

Taking steel bars at £3 17s. 6d. per ton, delivered free to the manufacturer's works; coal and slack, oil, grease and water; day labour to drive two or three mills, as millwright, engineer, stokers, coal-wheelers and labourers; tonnage labour, stocktakers, furnace-men, rollers, markers and shearers, and night foreman; contingencies; depreciation of rolls, breakages and repairs, I estimate the sheets can be produced from steel bars at £5 15s. for Doubles, and £6 5s. for Lattens, per ton, at the manufacturer's works, exclusive of dead charges, and upon these figures I base the cost price of producing galvanized corrugated sheets.

Staffordshire iron sheets are quoted to-day at £6 10s. for Doubles, and £7 for Lattens, per ton, at manufacturer's works, and this of course includes the manufacturer's profit.

The consumption of fuel naturally depends on the economical manner in which the boilers are constructed and fixed, and, as mentioned in another part, if the boilers are laid down at a great distance from the engine, and the pipes are uncovered, it can be expected that a considerably greater consumption of fuel must take place. The way in which the plant is arranged is, therefore, a question of profit and loss, and I can only assume that an intelligent manufacturer would take advantage of every possible means to economise, not only fuel, but labour also.

Approximate Estimate of Cost of Producing Galvanized Corrugated Sheets, Usual Sizes.

May 1, 1896.

	Doubles to 24 W.G.	Lattens to 26 W.G.
Sheets	1995 lbs. at £5 15s.	1890 lbs. at £6 5s.
Spelter	280 lbs. at £16.	360 lbs. at £16.
	£ s. d.	£ s. d.
Materials—		
Steel Sheets, Spelter	7 2 7	7 16 10
Acid, Ammonia	0 7 3	0 9 3
Slack, Coke, Water, Gas, Oil	0 2 7	0 3 0
Labour—		
Pickling, Dipping, Corrugating, Packing	0 7 0	0 9 3
Carriage to Liverpool (by Water from Swansea)	0 8 0	0 8 0
Sundries—		
Annealing and Fuel	0 3 0	0 3 0
Depreciation, Bath, Rolls, Tools; Difference in Discounts; Commission	0 7 6	0 8 8
Gross Cost per ton	8 17 11	9 18 0
Present lowest selling prices, f.o.b. Liverpool	9 17 6	11 7 6
Gross Profit per ton	0 19 7	1 9 6

Average for production, Half Doubles and Lattens, 24s. 6d. per ton, exclusive of Dead Charges.

HEATHFIELD'S PATENT PROCESS.

This machine consists of a pair of rolls at the front of the bath to take the sheets into the metal, and another pair at the other end through which the sheets emerge as they are coated. Both sets of rolls must necessarily be driven at the same speed. Between these rolls a guide is fixed, to guide the sheets through the metal. The rolls of the latter pair are placed in such a position that the "bite" of the rolls appears just above the surface of the metal, and the frames are fitted with levers and weights to regulate the pressure. In front of the inlet rolls a flux-box is fixed, but the sheets emerge through the clear metal, and do not therefore go into the water afterwards. It is claimed by the patentee that the pressure of the levers on the rolls ensures a thinner coating than by any other process, and less labour is required, as they are not washed, dried and brushed afterwards. It must be apparent, however, that the effectiveness of this apparatus, as regards the amount of metal in the coating, must depend upon the rolls being always so carefully regulated that the "bite" must be above the surface of the metal. The advantages I give in the patentee's own words. I shall be pleased to furnish a drawing showing the machine in operation. The size of bath used by the patentee for this process is 6 feet by 6 feet, by 2 feet 8 inches deep, inside measure, and costs about £45, and the patent machine £45. The royalty is 1*s.* per ton on the first fifty tons, and 6*d.* per ton afterwards per week.

The advantages claimed by the patentee are as follows :—

One man and one boy are sufficient to run the machine, and their combined labour will turn out far more galvanized sheets per day than a much larger number of men can do by any other process.

The machine will turn out with sufficient labour 12 tons or more of galvanized sheets assorted, 14 to 30 gauge or thinner, per turn of 10½ hours, and with a good dipper will sometimes do as much as 11 or 12 tons per turn of sheets assorted 26 to 30 gauge; while 8 English tons is a very moderate average in a turn of 10½ hours for sheets of these latter thicknesses.

The machine will do any thickness from 14 to 30 gauge or thinner, any width up to 4 feet (or wider if a large enough machine is supplied), and any required length.

As the pot is only 2 feet 8 inches deep, and holds only about 12 tons of spelter (zinc), the very large output, compared to the quantity of spelter in a molten state, reduces the dross made per ton of sheet below that made by any other plan; the pot of course being pushed to its maximum, and worked night and day.

No oxide is made while the machine is at work.

The quantity of muriate of ammonia used per ton of sheets is very small. Fourteen pounds, or less, will suffice to galvanize one ton of sheets, 28 gauge thick, while stronger sheets take proportionately less.

The coke bill is small. If the pot is kept fully at work, six tons of ordinary gas coke of fair quality will keep the pot at work a week, including firing it on Sundays.

The quantity of spelter used per ton of galvanized sheets made is greatly reduced, while the appearance of the sheets is much improved, and they leave the machine ready for use, without any subsequent brushing, washing, or drying.

The following are the approximate quantities of spelter deposited on the sheets, in the manufacture of one ton of galvanized iron by the Heathfield patent machine; the various gauges being calculated at the number of galvanized sheets per ton, which it is usual to supply in England:—

16	18	20	24	26	28	30	gauge
120	140	195	245	325	350	420	lbs.

In the very light gauges it is possible to use sheets a gauge stronger than can be used in the ordinary process, and yet produce the same number of galvanised sheets per ton as is customary by the old plan.

This saves considerably the cost of black sheets on 28, 29, and 30 gauge galvanized.

THE BAYLISS PATENT PROCESS.

The patentee claims that this method dispenses with a considerable amount of manual labour. The patentee passes the sheets, after being pickled, through a pair of cold rolls upon which a stream of water is continually flowing. The object of this is to impart a fine smooth surface to the sheets which have been roughened by the pickling process. From the cold rolls the sheet passes onwards towards the bath which it enters through a pair of rolls fixed on the brickwork of the galvanizing bath. It then passes through a guide fixed below the surface of the metal, and finally emerges through a layer of sand on the surface. The sheet is then seized by a pair of rolls having studs inserted at intervals which meet and grip the sheet. The sheet then passes on by means of an endless chain band to a set of revolving brushes which brush off any adhering particles of sand.

CARASCO'S PATENT MACHINE.

By this process the sheets pass through a pair of rolls on the bath, and thence by means of guides into the flux box. From the flux box they pass through guides in the metal, and emerge through a pair of wheels with V faces which grip the sheets by the edges. No provision is made for a flux box on the exit side of the bath.

ORIGINAL PROCESS.

The term "galvanized iron" is that which has for many years been given to articles of iron when coated with zinc. The object of such coating being to preserve the iron from oxidation by the atmosphere. When iron has been thoroughly cleaned and freed from scale it will, if dipped in a bath of molten zinc, become perfectly coated. When iron is properly coated, the atmosphere has no direct action on the iron, but a thin film of oxide is formed on the zinc coating, which is sufficiently hard to resist further oxidation and to remain perfectly sound. Galvanized sheet is generally corrugated, the wave-like form being given by pressing in dies. The great strength and stiffness which the corrugations give to the iron allow it to be used in sheets of considerable size for the roofs of buildings of small span almost without framework. The quality of galvanized sheet iron depends primarily on the quality of the black sheet iron, and then on the care with which it is coated with the zinc. *Good* galvanized sheets can only be made from a *good* quality of sheet iron, for if the iron is not cleaned and free from cinder it will not hold the zinc coating properly, and spots of rust appear which, eating into the sheet, make holes on the surface, and so the iron is exposed to oxidation and destruction. If the iron is not sufficiently tough and ductile, it will crack when corrugated, and, though the openings may be so small as to escape cursory examination, they will, when exposed to the weather, rapidly become rusty and render the whole sheet worthless.

The quality also depends on the degree of purity possessed by the spelter. Spelter is the name given in commerce to zinc before it is manufactured into sheets. The competition which has prevailed during the last twenty years in this, as in other trades, and the tendency to judge everything by the standard of price alone, have resulted in the production of a large quantity of very inferior galvanized sheets; and to this cause in no small degree is owing the disfavour with which galvanized iron is regarded by many who have used it. *There is no case where cheapness obtained by inferior quality is more a false economy than with galvanized iron, for the deterioration, once commenced, is so rapid as to be out of all proportion to the saving which may have been effected in the price at the first.*

When this trade was first established the sheets were universally galvanized by the "dipping process," no machinery whatever being used. The bath was generally 8 feet 6 inches, or 9 feet 6 inches, long by 2 feet wide by 4 feet deep, and would contain twice the amount of spelter of the baths of to-day. A bar of T-iron, upon which an iron plate was riveted, was placed on the bath longitudinally. This plate was just deep enough to go into the metal, when the bath was at its lowest working height. Its object was to divide the flux, so as to keep the flux on the exit side in the best condition. A dipper and under-hand were employed at the front of the bath, who jointly plunged the sheet into the metal and by means of their rods passed it under the bar, bringing it up through the flux on the other side. It was then seized by tongs held by the "takers-out," of whom two were employed, who gradually drew the sheet out of the metal, and when the surface was "set," or crystallized, plunged it into the water, it being passed afterwards to the sawdust boxes to be dried. If the sheet was required to be "bright" galvanized

(i. e. not crystallized), it was plunged into the water before the crystals had begun to form.

If the sheets were large, the two "takers-out" would be employed in the operation, and if the sheets were longer than the bath they would be doubled. By this process it was requisite that the sheets were carefully dried before entering the bath, as the least damp would cause the metal to fly. The process is objectionable in every way, as the coating is thicker and lacking in uniformity, and it is entirely dependent upon the skill of the men employed in taking out. In addition, there is a great consumption of muriate of ammonia, as the whole surface of the bath requires to be covered. The sheets are also liable to damage by reason of the accumulation of the dross at the bottom of the bath. When the dross is high the sheets have to be forced through it, often causing damage to thin sheets. This process for sheet galvanizing is now entirely dispensed with amongst English galvanizers, but I have seen it at work during recent years by some Continental makers.

Sand instead of Ammonia.—With a view to reduce the consumption of ammonia, it was often the practice to cover the surface of the metal on the exit side of the bath with black foundry sand. This was a reprehensible practice, and it is apparent to anyone that sand is not a flux, and its only imaginary economy would be that, from its dragging on the surface of the sheet, it would answer the same purpose as a flux. It made a large quantity of "sand skimmings" and oxide; and it is a difficult matter to effectually separate the sand and spelter when the latter is in small particles.

This is the system adopted by one patentee, and it shows a woful ignorance of the efficiency of ammonia as a flux, when this system is gravely put forth as an *economical method* superior to any other!

The Best Process.—The process in vogue by firms who produce a good quality economically, is as follows:—A strong, square iron frame is made, into which two rollers work. These rolls are made of the best hammered forgings, and are turned. The usual size is 3 feet 6 inches long, by 9 or 10 inches diameter. This frame and rolls are suspended from bars placed over the bath, and at such a depth that the rolls are completely immersed in the metal, and the "bite" of the rolls 14 to 15 inches below the surface. The rolls have a hammered wrought-iron pinion on the end of one, and two wrought-iron pinions at the other end, and are driven by another wrought-iron cog-wheel on a shaft placed on the side on the bath casing. The pinions and cog-wheel should be cut with teeth as large as practicable. The shaft has a fly-wheel and a four-cone pulley, and a corresponding cone-pulley on the overhead motion for varying the speed according to the thickness of the sheets. A thin sheet takes less time to coat than a thick one. It is advisable (though not universally adopted) to have a pair of rolls of 6 inches diameter at the entrance side of the bath to guide the sheets into the flux-box, its only object being to relieve the labour of the dipper. A flux-box is placed on the entrance side and one also on the exit side of the bath. From these flux-boxes simple guides are arranged, which are removed when the day's work is finished. Every competent dipper knows how these should be arranged, and he generally places them according to his own ideas of working. The sheet goes into the first flux-box in the wet state, and as it emerges from the flux-box on the other side it is seized by a boy armed with a pair of self-acting tongs, which are attached to a rope running over a pulley. The boy raises the sheet in a vertical position, and, when quite clear of the metal, by an adroit movement he rolls it up and plunges it into the water. A mechanical movement can be arranged with a clutch to wind the rope, and so ease the boy's labour. Two water "boshes" are used, placed side by side, and as the water in the first gets hot rapidly, the sheets are plunged into the second, to prevent them from drying white before they are sawdusted. In addition to this the hot water readily dissolves any particles of flux that may have adhered to the surface of the sheet. From the water they are passed to girls, who rapidly brush them in a trough containing sawdust, finally drying them singly over an open coke fire. It is essential to pass the sheets through a pair of 6 inch rolls after leaving the bath to ensure the corners and edges being level preparatory to the corrugating process. This process is the best, and the one I worked out so successfully for many years when engaged in the manufacture.

The action of the rolls in the bath keeps up a constant circulation of the metal, and the dross is not so likely to get solidified as by any other process. But the chief feature is the flux-box at the exit side of the bath. It is an acknowledged fact that the muriate of ammonia *liquefies* the metal, which can easily be proved by dipping a small article and withdrawing it through the clear metal, and dipping one and withdrawing it through the flux. The action of muriate of ammonia can also be readily seen when a bath is filled, and before the machinery is introduced and no flux is on the surface. A very slight film of oxide, varying according to the heat of the metal, will be seen floating on the surface in all the colours of the rainbow. By scattering a very small quantity of the muriate over the surface, this film will instantly disappear, leaving the metal on the surface as bright as molten silver. This is a sure proof of the economy in the use of a flux, in contrast to any other method.

Care in Drying.—Too much care cannot be exercised in the drying of the sheets after the sawdusting process; and to ensure this being effectually done it is quite as requisite for the night watchman to have the drying fires in order when the work *commences* as it is for the galvanizing bath to be in order. If the drying is carelessly performed, the result will be that the sheets will "sweat" after packing, large patches of oxide appearing on them, and the longer they remain packed the worse will be the effects; so that a manufacturer's responsibility does not end at the shipment of the goods, but will follow him until they are unpacked for distribution. While residing in Valparaiso, Chili, I was called upon to decide a case of this kind. The sheets that I inspected were very well galvanized, and the packing was perfect. As the sheets were of small corrugations, for the fronts of houses, they were packed in skeleton cases, as it is not practicable to bundle sheets with small corrugations without damaging the outside edges. Apparently the sheets were faultless, but on opening the cases large patches of oxide, varying from 12 to 24 inches in diameter, were observable on nearly every sheet. One damp sheet will readily damage the next, even if a dry one. I unhesitatingly came to the conclusion that the damage arose through insufficient drying. If the damage had been caused on the voyage the action of the sea-water would be observed on the edges of the sheets, and more so than in the middle. The result was, that the manufacturer was mulcted in a heavy claim for damages. When this occurs it is not lightly done by any means, and a manufacturer who has experienced it once receives a lesson he is not likely to forget!

WEIGHT OF COATING.

The weight of spelter taken up in coating is an average, and can easily be tested by taking the gain per sheet. All sheets for galvanizing are rolled *light*, and taking a sheet 6 feet by 30 inches, by 24 gauge (exact width for 8-3 inch corrugations is $29\frac{1}{2}$ inches), containing 15 square feet, it would be, by the B.W.G., 16 ounces per square foot, or 15 lbs. per sheet. For galvanizing these would be rolled $14\frac{1}{4}$ lbs. per sheet. When galvanised these should run according to the standard of weights in the galvanized iron trade, about 140 sheets per ton; so taking

```
                                          lbs.
  140 sheets, at 14¼ lbs. per sheet  =   1995
  Coating 140 sheets, at 2 lbs. each =    280
                                         ————     ton  qr.  lbs.
                                         2275 =    1    1    7
```

This is a gain in spelter of about 15 per cent. By Heathfield's patent process it is professed to coat with 235 lbs. per ton, but it is quite possible to put too little on!

The number of sheets of the various lengths and gauges is given in the Table of "Standard of Weights in the Galvanized Iron Trade," but the relative number of the various lengths can always be ascertained by taking the six-feet sheets as a basis. For instance, in the above, 140 sheets 6 feet long equal 840 lineal feet, and this divided by 5 feet gives 168 sheets, and by 7 feet gives 120 sheets, and by 8 feet gives 105 sheets, and by 9 feet 93 sheets, and so on in proportion for larger and intermediate sizes.

We will now take 26 w.g., which for a sheet 6 feet by 30 inches, by 26 gauge, would be rolled $10\frac{1}{2}$ lbs. for galvanizing; so taking

```
                                             lbs.
180 sheets, 6 ft. by 30 in. by 26 ⎫
    w.g., at 10½ lbs.      ..  .. ⎭ = 1890
Coating 180 sheets, at 2 lbs. each =  360
                                     ─────    ton  lbs.
                                     2250 =   1    10
```

This shows a gain in spelter of nearly 20 per cent., and for what would be sold as actual 26 gauge. With Heathfield's patent machine it is professed to coat with 300 lbs. per ton, or about 17 per cent.

For the Australian markets the gauges sold as 26 w.g. are actually light 27 w.g. in the black, and take up more in coating, but a relatively better price can be obtained. They should weigh in the black barely 10 lbs. for a 6-feet sheet, and other lengths in proportion.

CLOSE ANNEALING.

The sheets as they leave the rolling-mill, being in the "hard" state, require softening before being pickled. When the old process of dipping was in vogue it was the custom to soften the sheets in an open furnace; but by this method the sheets are not softened uniformly, and are also liable to be burnt on the edges. The practice of "scaling" the sheets then came into operation. This was effected by first placing them in a tank of old acid and afterwards conveying them, two or three at a time, according to the width, into an open furnace. The sheets rested on triangular fire-brick bearers placed on the bed of the furnace. After an exposure to the heat of two or three minutes they were taken out and placed on a cast-iron bench, perforated with holes, and beaten with broad hoop-iron beaters. This process effectually removes the scale, but it is not only slow but expensive, as a badly regulated furnace will cause more scale to rise than is necessary, thus involving loss of material. This system has now been more or less abandoned, but one scaling furnace is a useful adjunct to a galvanizing works, in case the annealing furnaces are out of order.

Close Annealing.—This was formerly done with the "old style" of furnace, with one door and one fire-hole at the side, using coal. The boxes, holding 5 to 6 tons at each charge, were conveyed into the furnace on cast-iron rollers.

The new type of furnace has a door at each end, and is fired at the side by three or four fire-holes, according to the size of furnace. The fires are blown with a steam jet placed under the grate in each fire-hole. Only very small coal, or slack, is used for the fires. The bed of the furnace has a pair of cast-iron plates, with a V groove in each, and the annealing box rests on cast-iron balls, which move in these grooves, and it is easily propelled in and out of the furnace. The box is filled at one end, moved into the furnace, and, when the annealing is completed, it is taken out at the other end, so that a box can be filled while one is in the furnace, and so save considerable time in moving boxes out of the way, as was the case with the old method.

ANNEALING BOXES AND FURNACES.

The boxes hold from 12 to 15 tons at each charge. The heat being more uniformly distributed than by the old method, the boxes are not destroyed so quickly. The estimated cost of annealing by this method, including labour, fuel, renewal of boxes, repairs of furnace, etc., is 2s. 9d. to 3s. per ton. Where the output of the galvanizing works requires the use of two furnaces it is more economical to build a pair, as the respective cost is not so great.

The approximate cost of a double annealing furnace, to take in 10-feet 6-inch boxes, would be £450, including roof and overhead traveller.

The covers of the boxes should be of wrought iron, 1 inch thick, and the following are the most useful sizes.

For sheets up to 8 feet long by 30 inches wide: 8 feet 6 inches by 3 feet, by 3 feet 6 inches deep, inside measure.

For sheets up to 10 feet long by 30 inches wide: 10 feet 6 inches by 3 feet, by 3 feet 6 inches deep, inside measure.

The approximate weights are 48 cwt. and 60 cwt., respectively.

The dishes will do of either cast or wrought iron, and the approximate weights in the latter material are 55 cwt. and 70 cwt. respectively.

It is advisable to have covers and dishes of both sizes. The 8-feet 6-inch will do the *bulk* of the corrugated sheets required, and up to 2 feet 9 inches wide. The 10-feet 6-inch box should be kept for 9 and 10-feet sheets, as it is not advisable to anneal 6-feet sheets in a 10-feet 6-inch box, as the top of the cover is liable to "sag" if unsupported.

For flat sheets the covers should be 6 feet 6 inches by 3 feet 6 inches, by 3 feet 6 inches deep, to take sheets up to 3 feet wide.

PICKLING.

Not knowing whether muriatic acid can be readily procured in the neighbourhood of the works, or within a reasonable distance, in South Wales, I have assumed that sulphuric acid would be used. Muriatic acid is generally used by the galvanizers, but the relative cost is about the same, as though sulphuric acid is about double the price of muriatic acid, it takes less. Muriatic acid has this advantage, that the pickling can be performed quicker, and the sheets galvanized straight away from the acid, whilst pickling with sulphuric acid not only takes more time, but the sheets must stand in water for some time preparatory to being galvanized.

The gravity of sulphuric acid for pickling should be 1·850. Proportion, five parts of water and one part acid. Time required, about 20 minutes. Duration in water-tanks, 24 hours. Sulphuric acid weighs 19 lbs. per gallon, so that the weight required in a tank can be easily ascertained by taking the cubical contents. The marketable value is about £3 10s. per ton.

Muriatic acid is generally of the strength of 35 Twaddle, but this strength is not guaranteed, it being made as a waste product. The proportion is generally half acid and half water. Time required, 12 to 15 minutes. Muriatic acid weighs 19½ lbs. per gallon, and the market value is £1 15s. per ton. Wood tanks lined with stout sheet lead will answer as pickling tanks for sulphuric acid. A removable bottom, of 3-inch elm plank, should be wedged in, to prevent the sheets from cutting the lead. A wooden rim, standing a little over the edge of the tank, should go all round the tank to preserve the top edges of the lead. A steam-pipe of lead, for heating the acid to about 90°, should be inserted in one corner of the tank, and within three inches of the bottom. The pickling tanks for muriatic acid require to be made of Yorkshire stone, with solid round rubber jointing, and clamped with cast-iron corner clamps and stout iron rods, which should be well tarred at frequent intervals. The fumes from pickling with muriatic acid are not so distressing to the workmen as the fumes from pickling with sulphuric acid, and, where it is possible, it is highly essential to use the former.

The sheets enter the galvanizing bath in a wet state, and it is the practice at many works to pass the sheets direct from the pickling tank (after muriatic pickling), or from the water tanks (after sulphuric acid pickling), with only a slight draining, direct to the galvanizing bath. A considerable amount of surplus moisture can be taken up by using a flat shallow tank which is placed in a line with the galvanizing bath. At one end of this tank (which should be substantially constructed) a pair of rolls are fixed which are driven by a band from the same overhead shaft driving the galvanizing machinery. These rolls are of wrought iron, covered with indiarubber, and 6 inches diameter by 4 feet long. A strong spring at each end is sufficient to give all the pressure required. The sheets are thrown into this tank from the pickling or water tanks, and passed singly through these rolls as required by the dipper. A table can be placed in front of the rolls for the sheets to drop upon, or if rolls are used on the entrance side of the bath, a guide can be arranged to conduct the sheet from one set of rolls to the other. Nothing is gained by the use of such a guide, and I prefer a dipper to start the sheets into the flux box, as guides and rolls on the bath are in my opinion only an obstruction.

Muriatic acid is conveyed in glass carboys, supplied free by the makers of the acid. These, being made of blown glass, vary very much in thickness, from $\frac{1}{4}$ inch to $\frac{1}{16}$th inch thick, and are, consequently very liable to breakage. This is occasioned very often through the bad packing of the carboy in the wood or iron frame, causing the carboy to fall out in the process of emptying, and also from exposure to frost. It is, therefore, desirable to keep the stock of muriatic acid under cover, in the immediate vicinity of the pickling tanks, seeing also that the earthenware corks are air-tight with clay, as this acid has a natural affinity for moisture, which quickly reduces the strength. It is a waste product, so the makers will not guarantee the strength. When the requirements of the works are known it should be ordered only in such regular quantities as are required for a week's consumption, at the same time ensuring a little stock, in case of delay in transit. The carboys are supplied free, and it should be clearly understood that, while every care is taken in the handling of them, no responsibility whatever will be incurred for breakages. I have known cases where the supplier has endeavoured to obtain a receipt for the safe custody of *all* the carboys by supplying a Dr. and Cr. account every month, showing the balance due to him, which of course included breakages, and he eventually holding the galvanizer liable for every carboy missing! I suppose more breakages occur from the irregular thickness of the glass of these carboys than from any other cause. Responsibility should be totally repudiated from the commencement, and the receipt should be given in such a form that it is a receipt for the contents only.

My figures are based on pickling sheet iron, which carries considerably more scale than steel, and the quantity can be taken as the maximum. It can be gauged at 1¼ cwt. per ton of sheets, in sulphuric acid, and 2½ cwt. per ton with muriatic acid, for doubles, and lattens *pro rata*. Those figures should be checked by keeping an account of the acid consumed, and the production of galvanized iron, for a month.

The economy in the quantity used, forms one of the chief characteristics of a competent head pickler, and it can be safely assumed, that a careless and incompetent man will use considerably more for the same quantity of work.

PRICES FOR LABOUR:
PICKLING, DIPPING, CORRUGATING, PACKING.

Labour.—This varies with the different works, some paying by tonnage, and others paying by the count. The rates per ton are usually as follows :— pickling, 1s. 8d. ; dipping, 2s. ; corrugating, 1s. 6d. ; packing, 1s. 6d. per ton. This involves a considerable amount of labour in weighing the sheets, and, as the actual results in paying by weight and paying by count are, that in the former case you pay on the *galvanized* article whereas by count they are paid on the sheets as supplied to each pickling tank, the thinner the sheets are, a proportionally larger quantity can be done, and *vice versa*. There are no uniform prices amongst the workmen, neither is there a Trade Union, so that it rests on the manufacturers to arrange the prices according to the competency of the men employed. The principal responsibility of galvanizing rests with the dipper, and his skill consists in keeping the metal at the proper working heat—and, in short, the whole care of the bath devolves upon him. For paying by count the following can be taken as a fair wage; and presuming your output of sheets would be such as to keep the baths fully employed, I should recommend paying the pickling and dipping by count, as sent into the galvanizing shop, and this is the cheapest and readiest plan. The corrugating and packing can be paid by the ton, according to the weight packed, only settling for this weight according to the weigher's book. Supposing that the output of sheets is such as to keep a bath fully employed, the following can be taken as a fair

OUTPUT FROM ONE BATH.

	18 & 20	22 & 24	26	28 & 30	w.g.
Weight per week-days only	50	50	45	40	tons.
Number of sheets in—					
5 & 6 feet, about	4200	7200	8400	9600	sheets.
7 ,, ,,	3600	6200	7200	8200	,,
8 ,, ,,	3150	5400	6300	7200	,,
9 ,, ,,	2800	4800	5600	6400	,,
10 ,, ,,	2520	4320	5000	5800	,,

PICKLING AND DIPPING LABOUR PER HUNDRED SHEETS.

The following may be taken as a fair average basis of labour by count:—

PICKLING (including pickler and his under-hand).

	18 & 20	22 & 24	26	28 & 30	w.g.
	s. d.	s. d.	s. d.	s. d.	
5 & 6 feet	2 6	.. 1 6	.. 1 3	.. 1 0	per 100.
7 ,,	3 0	.. 1 9	.. 1 6	.. 1 3	,,
8 ,,	3 6	.. 2 0	.. 1 9	.. 1 6	,,
9 ,,	4 0	.. 2 3	.. 2 0	.. 1 9	,,
10 ,,	4 6	.. 2 6	.. 2 3	.. 2 0	,,

DIPPING (including dipper and his under-hands).

	18 & 20	22 & 24	26	28 & 30	w.g.
	s. d.	s. d.	s. d.	s. d.	
5 & 6 feet	3 0	.. 1 9	.. 1 6	.. 1 3	per 100.
7 ,,	3 6	.. 2 0	.. 1 9	.. 1 6	,,
8 ,,	4 0	.. 2 3	.. 2 0	.. 1 9	,,
9 ,,	4 6	.. 2 6	.. 2 3	.. 2 0	,,
10 ,,	5 0	.. 2 9	.. 2 6	.. 2 3	,,

Above is for 30 inches, or 29½ inches wide, for 8-3 inch corrugations. A proportionate increase should be paid for wider widths.

CORRUGATING (including corrugator and under-hand).

	18, 20, 22 & 24	26	28	w.g.
	s. d.	s. d.	s. d.	
5, 6, 7 & 8 feet	1 6	.. 1 9	.. 2 0	per ton.
9 ,,	1 9	.. 2 0	.. 2 3	,,
10 ,,	2 0	.. 2 3	.. 2 6	,,

Above is for corrugating, all widths.

PACKING.

All sizes and gauges, including under-hands, 1s. 6d. per ton.

Settlements for corrugating and packing to be made according to weigher's book.

PRICES FOR GALVANIZED CORRUGATED SHEETS.

May 1, 1896.

Usual lengths.	Best, 8-3 inch corrugations wide. £ s. d.		Best Best, 8-3 inch corrugations wide. £ s. d.	
18 to 20 w.g. ..	9 12 6	per ton. ..	10 2 6	per ton.
22 to 24 ,, ..	9 17 6	,, ..	10 7 6	,,
26 ,, ..	11 7 6	,, ..	11 17 6	,,
28 ,, ..	11 17 6	,, ..	12 7 6	,,
29 ,, ..	12 7 6	,, ..	12 17 6	,,
30 ,, ..	13 0 0	,, ..	13 10 0	,,

Above are the lowest current prices quoted on this date, including packing, in bundles, and delivered f.o.b. Liverpool in 10 ton lots and upwards. Lots under 10 tons, but over 2 tons, 2s. 6d. per ton extra. For lots under 2 tons extra carriage is charged. If taken at Staffordshire Works the prices are 10s. per ton less. Terms for cash against shipment are usually 3 per cent., or 2½ per cent. for cash tenth month following delivery.

What is Best and Best Best? It will be noticed that there are two lines of prices, which require some explanation, and also support my theory (as explained under another heading) as to the necessity of knowing the requirements of the markets. It will be noticed that there is a difference of ten shillings per ton, between the "Best," and the "Best Best" qualities.

After an experience of twenty-five years I should be unable to enlighten a buyer who asked the question as to wherein the difference consists! In plain English, there is no difference at all, except in the price. It is only a favourite and peculiar method.

with some manufacturers, who try to court the requirements of every one with the lowest prices on the one hand, and an additional profit on the other. My theory is (as explained elsewhere at greater length), for the sake of economy in the manufacture it is requisite to have a sheet with a good surface; and when this fact is accepted as being absolutely correct, it also follows that it is equally as cheap to coat a sheet well, as to coat it badly. A manufacturer's aim should therefore be to manufacture one uniformly good quality, and he will soon be able to realise a uniform compensating remunerative price without having the necessity to dodge between the difference between "Best" and "Best Best," which is actually about the same as between "tweedledum" and "tweedle-dee"!

Assuming that you will not issue a printed Prices Current, I would suggest that you should put on your card and note-paper: "Manufacturers of Galvanized Corrugated Sheets of the best quality only," and not the "best" with a capital B. When you quote, you should quote for your brand, and the headline on your paper will convey the impression that you can only make one quality, and that of the best, and no further explanation will be required.

"*All round prices.*"—A practice has existed with buyers for the Indian Market, to ask for an "all round," or average price for all the gauges in a specification—say, from 18 to 24 w.g.—to save expense in cabling. As the average will come out according to the specification, it follows, if the singles preponderate, the average will be less than the price for doubles, thereby enabling the buyer to say he can buy doubles at a lower price than he actually can do if the specification was all doubles.

It is only a matter of figures, and the best plan is to let the buyer find the average out for himself and *not* the manufacturer, so that no ground can be left him for saying "that he is actually quoted by a manufacturer."

EXTRAS ON GALVANIZED CORRUGATED SHEETS.

The usual extras on lengths, etc. are as follows :—

18 to 24 w.g.—8-3 inch corrugations: 10 feet, 10*s.*; 11 feet, 30*s.*; 12 feet, 50*s.* per ton.
    10-3 inch corrugations: 9 feet, 10*s.*; 10 feet, 20*s.*; 11 feet, 40*s.*; 12 feet, 60*s.* per ton.
    11-3 inch corrugations: 6 to 8 feet, 20*s.*; 9 feet, 40*s.*; 10 feet, 60*s.* per ton.

26 w.g.—8-3 inch corrugations: 9 feet, 10*s.*; 10 feet, 20*s.*; 11 feet, 40*s.*; 12 feet, 60*s.* per ton.
    10-3 inch corrugations: 9 feet, 20*s.*; 10 feet, 40*s.*: 11 feet, 60*s.*; 12 feet, 80*s.* per ton.

28 w.g.—8-3 inch corrugations: 9 feet, 10*s.*; 10 feet, 20*s.* per ton.
    10-3 inch corrugations: 9 feet, 20*s.*; 10 feet, 40*s.* per ton.

29 w.g.—8-3 inch corrugations: 9 feet, 20*s.*; 10 feet, 40*s.* per ton.

30 w.g. not supplied over 9 feet long. Lengths over 8 feet are extra, viz. 9 feet, 30*s.* per ton.

Special quotations for 2-, 4- and 5-inch corrugations. Curving sheets to any ordinary radius, 10*s.* per ton. Ordinary punching one side and one end, 10*s.* per ton.

PRICES OF GALVANIZED FLAT SHEETS.

May 1, 1896.

6 feet by 2 to 3 feet wide ..	18, 20	22, 24	24	28 w.g.
	per ton. £ s.	per ton. £ s.	per ton. £ s.	per ton. £ s.
(1) Best, close annealed ..	10 10	10 15	12 5	12 15
(2) Best patent flattened, cold rolled & close annealed	12 15	13 0	14 0	15 0
(3) Best Best ditto	13 15	14 0	15 0	16 0
(4) Best Best Best ditto	15 15	16 0	17 0	18 0
(5) Charcoal ditto	23 0	23 5	24 5	25 5
(6) Soft steel ditto	15 0	15 5	16 15	17 15

Above is the copy of the current price list of a galvanized iron manufacturer, and furnishes another example of the unnecessary complications resorted to by some makers in the trade. If the manufacturer is not puzzled himself he certainly must puzzle the buyer who is so unfortunate as to have the task to select what he requires from the six different descriptions and lines of prices!

There are three stock sizes of flat galvanized sheets—viz. 72 inches by 24, 30, 36 inches wide, so that to be able to supply any of the above quickly, a manufacturer needs to have 72 different sizes and qualities in stock, or in process of manufacture, and this is a fairly good proof of the absurdity of the whole thing!

Before going into full particulars of what is actually required, it will be as well to inquire—

WHAT ARE PATENT FLATTENED SHEETS?

On referring to the above price list it will be noticed that the sheets on the second line are described as Patent Flattened, Cold Rolled, etc. The term arose through the use of Britton's Patent Levelling Machine in connection with sheet iron. I have not much knowledge of this machine, but consider that cold rolling answers the purpose better. If a sheet is flat the buyer does not trouble himself by what process this end has been obtained. The term was originated by a galvanizing firm who at one time had a mania for attaching the word "patent" to everything they manufactured with the idea of attaching an importance to the article not possessed by any other makers. It has been adopted by one or more firms besides, who have added the words "cold rolled." If the sheets are "patent flattened" and *are* flat, they do not require to be cold rolled, and if they *are* cold rolled they do not require to be "patent flattened." The manufacture of galvanized flat sheets is considerably more lucrative than that of galvanized corrugated sheets. It is, however, a great mistake to pretend to make a great variety of qualities for the reasons above described.

The first quality required is that of plain galvanized sheets of the ordinary roofing quality, and this corresponds with the first line of the above price list. These sheets are not cold rolled, but close annealed only. It will be seen that there is a difference in favour of the flat sheets of 17s. 6d. per ton, without taking into consideration that no corrugating is necessary. Even assuming that they were sold at the same figure as corrugated sheets, the price would leave a profit over the latter. But this is not the case, for 5s. to 13s. per ton can be realised over corrugated sheets for a good make of iron or steel in the thin gauges for the Australian and New Zealand markets. This iron will not work up, and this should be stated in the quotation, but 24, 26 and 28 w.g., if it is well annealed, is quite good enough for making into spouting and ridging, and there is a very large demand for these purposes for the Australian and New Zealand markets, and the sizes required are 72 inches by 24, 30, 36 inches, by 24, 26 and 28 gauges.

The next class of flat galvanized sheets required is of a superior quality that *will work up*, and these should be annealed and pickled, cold rolled, and again annealed. This should be a quality of iron equal to the fourth on the above list, and if the manufacturer does not make it himself, he can afford to buy it in the "hard" state from such makers as Knight and Crowther, Kidderminster, who make this quality of black sheets as their "M" brand.

The next quality should be equal to any work that may be required to be put upon it in *all* gauges, and this iron can be bought from the same makers.

It is as well to test the iron after galvanizing in each quality. The first should stand bending the full length of the sheet, but not with an acute edge. As you will use it for your own ridging and gutters, it will be easy to ascertain that it will be suitable for others.

The next class of iron should be good enough to wire and groove in 20 w.g. and thinner.

And the third class should be good enough to stand doubling twice in all gauges, without breaking at the corners. One essential is, that the coating does not peel off at the edges after bending, and for this reason I am not in favour of steel sheets for working up purposes.

For the latter two qualities a fully compensating profit can always be realised, and the relative cost can easily be obtained in comparing with that of galvanized corrugated sheets, and for this reason the quality should be the *primary* consideration.

STANDARD OF WEIGHTS.

The accompanying table (see overleaf) shows the usual number of sheets per ton of the respective sizes and gauges.

Australian Buyers generally specify the number of sheets per ton, or per case, and they often run even more in number than that here given. The sheets required for the Australian Market, although marked as 26 w.g., are actually light 27 w.g., or nearer 28 w.g. in the black state. Great care is requisite in supplying the requirements of the market, as they are generally retailed there at per sheet, and a buyer depends on getting the specified number per case.

BLACK CORRUGATED SHEETS; CONTINUOUS ROOFING; ROOFING TILES.

Black Corrugated Sheets.—These are the sheets not galvanized, and are simply annealed and corrugated. The only labour is the corrugating and bundling, which has to be added to the current price of the sheets.

If painted corrugated sheets are required there are the three items of corrugating, painting and packing; the latter being generally in skeleton cases. The best kind of paint to use is the ready-

APPROXIMATE NUMBER OF GALVANIZED CORRUGATED SHEETS PER TON REQUIRED FOR EXPORT.

	ft. 5	ft. in. 5 6	ft. 6	ft. in. 6 6	ft. 7	ft. in. 7 6	ft. 8	ft. in. 8 6	ft. 9	ft. in. 9 6	ft. 10	ft. 11	ft. 12
16 B.G. 5-5 inch Flutes	70	64	58	54	50	47	44	41	39	37	35	32	29
,, 6-5 ,,	59	54	49	45	42	39	37	35	33	31	29	27	24
18 B.G. 5-5 ,,	86	78	72	66	62	57	54	51	48	45	43	39	36
,, 6-5 ,,	74	67	62	56	53	50	46	43	41	39	37	34	31
20 B.G. 8-3 ,,	114	104	95	88	81	76	71	67	63	60	57	52	47
,, 10-3 ,,	95	86	79	73	68	64	59	56	53	50	47	43	39
22 B.G. 8-3 ,,	139	127	116	107	99	93	87	82	77	73	69	63	58
,, 10-3 ,,	116	105	97	90	83	78	73	68	65	61	58	52	48
24 B.G. 7-3 ,,	189	172	157	146	135	126	118	110	104	99	94	85	78
,, 8-3 ,,	168	153	140	130	120	112	105	98	93	88	84	76	70
,, 9-3 ,,	154	140	128	119	110	103	96	90	85	81	77	70	64
,, 10-3 ,,	140	128	117	108	100	94	88	83	78	74	70	64	58
26 B.G. 7-3 ,,	251	228	209	193	178	167	156	147	139	131	125	..	..
,, 8-3 ,,	223	203	186	172	159	149	139	131	124	117	111	..	..
,, 9-3 ,,	204	186	170	157	146	136	127	120	113	107	101	..	..
,, 10-3 ,,	186	169	155	143	133	124	116	109	103	98	93	..	..
28 B.G. 7-3 ,,	270	246	225	208	193	181	168	158	149	142	135	..	..
,, 8-3 ,,	240	219	200	185	172	161	150	141	133	126	120	..	..
,, 9-3 ,,	220	200	183	169	158	147	137	129	122	116	110	..	..
,, 10-3 ,,	200	182	167	154	143	133	125	118	111	105	100	..	..
29 B.G. 7-3 ,,	296	271	247	228	212	198	184	174	164	155	148	..	..
,, 8-3 ,,	264	241	220	203	189	176	165	155	146	138	132	..	..
30 B.G. 7-3 ,,	324	297	270	249	231	216	202	191	180	..	..	..	..
,, 8-3 ,,	228	264	240	222	206	192	180	170	160	..	..	..	..

mixed Torbay paint, specially prepared for this class of work, costing about 28s. per cwt. Half a cwt. will coat 500 square yards, which is equal to one ton of corrugated sheets. The paint can be applied with a brush, by female labour, and will dry hard in the open air. Only one coat is required. It is essential to see that they are *quite dry* and hard before packing, or the compression of the packing will cause them to adhere together.

Continuous Roofing Sheets.—These are flat galvanized sheets, connected together in continuous lengths of 100 to 200 feet, as may be required. They are, as a rule, supplied only in the thin gauges, as 26, 28 and 31 gauge; the 31 gauge being 24 inches wide, and the others 24 or 30 inches, as may be required. They are rolled 6 to 9 feet long, according to convenience, as the length is of no consequence. They are annealed, and, after being pickled, are passed along a table, where they are edged and locked by a machine as fast as they travel into the bath. This roofing is packed in iron kegs, and enough 1-inch flat-headed galvanized nails to nail it down included in the kegs. The current prices are as follows:—

  24 inches wide, 31 w.g., $2\frac{1}{4}d.$ per square foot.
  24 or 30 inches wide, 28 w.g., $2\frac{1}{2}d.$ per square foot.
    „ „ 26 „ $2\frac{3}{4}d.$ „
  Kegs free. Nails 10d. per 1000.
  Delivered f.o.b. London or Liverpool.

The order should specify the width required, and nails are always sent, unless specified to the contrary.

Galvanized Iron Tiles.—These are made from flat galvanized sheets by pressing them between two dies, which can be fixed in the corrugating machine. They are rolled in a sheet large enough to make two tiles, and, after galvanizing, are cut through. The size is 36 inches by 24 inches, after being pressed. They have a roll, or corrugation, on each side, and four semicircular ribs in the centre, at equal distances. One roll is bolder than the other, to ensure them being laid equally, and this roll is punched with three holes. The price generally averages 5*s.* per ton above the price of galvanized corrugated sheets, with the usual extra for packing, which is generally skeleton cases. The following is the average number per ton for each gauge :—

    3 feet by 2 feet by 24 gauge, about 335 per ton.
        ,,   ,,   26   ,,   400 ,,
        ,,   ,,   28   ,,   460 ,,
        ,,   ,,   30   ,,   540 ,,

They are generally shipped to the West India Islands.

WHAT ARE GALVANIZED *TINNED* SHEETS?

The question is often asked, What is the difference between galvanized tinned sheets and galvanized sheets? Practically, there is no difference at all. In explaining the old process of galvanizing I showed how the "bright," or not crystallised, galvanized sheets were made, and this will explain what little difference there is between these two terms.

The spelter requires a little tin in it to produce the well-known crystallisations, or "spangles," on the sheets, although some spelter will actually crystallise without it. The V M spelter will galvanize "bright" unless tin is mixed with it, and this is the case with several other brands. The quantity of tin in the metal does not affect the coating so much as to warrant the term "tinned," but it can be generally understood that any maker using the term—although it is unnecessary, and not in general use—implies that his sheets *are* crystallised. I do not, however, know any maker who would venture to send them out otherwise, unless he was instructed to do so. I have known Government and Indian specifications to specify for sheets without the crystallisations.

TO ASCERTAIN THE QUALITY OF SPELTER.

To ascertain the quality, the plate of spelter must be broken through, and the fracture scrutinised. If the granular face is large and regular it shows the spelter is pure, but if small it is a proof that the spelter is not of first-rate quality. The quality of spelter is most frequently deteriorated with iron, and in very small particles. These being mixed with the spelter, separate it into so many separate fragments, the number depending on the amount of iron contained in the spelter. A simple illustration of this can be shown by taking some dry black sand and introducing into it some quicksilver, and well mixing them together; the quicksilver representing the spelter and the black sand the oxide of iron. Now if you take a *powerful* magnifying glass and examine a piece of spelter you will find black particles acting as *wedges* between the particles of spelter, which will clearly show the reason for the different sizes of these in various qualities of spelter. The more the particles of iron abound the finer the particles of spelter are separated. If the glass is applied to a piece of "dross" it will be found that the particles of iron abound still more freely, up to 70 or 80 per cent.

The term "virgin spelter" is applied to spelter made directly from the ore, and is of course the purest; if the dearest in price, it is the cheapest in the end. A large quantity of spelter is made from zinc cuttings and old zinc, and must evidently contain more or less impurities. Even the dross is manipulated, and sold under the name of "remelted" spelter, but the price is no criterion whatever of the quality. It is the cheapest plan to buy the best that can be procured, as the dross is made quite fast enough in the galvanizing process without introducing it through the spelter. Spelter has also some lead with it at times, but this will appear on the outside of the plate, as it will not mix, and when melted it always sinks to the bottom of the bath, and does not affect the metal in any way. The best English spelter is the "English crown" brand, and the best foreign spelters are S S, also P H, C G H, W H, and V M. The H brands are termed "specials," and V M is generally the dearest, as the Vieille Montagne Company use so much themselves for rolling into zinc that the supply is intermittent, and cannot be depended upon. It is, however, acknowledged as the purest of all brands.

GALVANIZING BATHS.

The duration of the galvanizing bath varies considerably, and sometimes leads to disputes between the maker of the bath and the galvanizer. One of the chief essentials to the duration of the bath is the quality of the iron or steel of which it is constructed. It should be made of the very best that can be obtained. The quality should be the *primary* consideration in determining to whom the order should be given. The make of material should also be determined upon and *specified*, and if each plate is branded, and the workmanship is perfect, the galvanizing bath maker cannot do more, and this should be considered as ending all responsibility on his part. I am strongly in favour of welded baths in preference to riveted ones, as a considerable amount is saved in the weight, as a welded bath will weigh about one-fifth less, and it will wear more equally than a riveted one. A riveted bath always wears away more quickly at the joints, through the plate being strained in the process of punching. Besides this, if a riveted bath is carelessly riveted (the caulking will hide this defect to a certain degree) and the joints are not close, the metal will get between the joints, and it is then impossible to make them sound afterwards.

If the manufacturer prefers a riveted bath he should specify that the holes are to be drilled, and the edges of the plates planed, and the joints well caulked inside and outside, and this will have a material influence on the soundness and endurance of the bath.

The best form for a welded bath is to have the two sides and bottom in *one* plate, and so save the exposure of welded seams to the fire. A useful size for sheets is 5 feet by 5 feet by 3 feet deep, made of 1-inch plates, *weighing about* 30 *cwt.*

Depreciation of the Bath.—The chief feature in the endurance of a bath is the manner in which they are fixed, so that the firing is equally performed, and, lastly, the freedom from undue draughts in the galvanizing shop, which will often force the fire unduly at one particular part of the bath. I have known baths worn out in nine months, while others have lasted two years, and this clearly shows the necessity of ascertaining and studying all the features upon which the existence of the bath depends. It is important also, in fixing the bath, to ensure it from warping, to have cast-iron vertical bars, 2 inches thick and the width of the flue, built in at regular intervals, and these should reach within 7 or 8 inches of the bottom of the flue.

The Endurance of the Rolls in the baths also depends on the quality of the iron used, which should be of the best hammered scrap forgings. Where there are two or more galvanizing baths it is desirable to have a 10-inch centre lathe on the premises, for turning the new rolls and re-turning the old ones, which requires to be done occasionally. A man who can work at the lathe and the fire is the most desirable man to have, as he can make and repair all the tools required in the galvanizing shop. The forgings can be procured from any manufacturers of good wrought-iron forgings.

STOPPING LEAKY BATHS.

It often occurs that a bath may be worn more in one place than in another, and this is often caused by extreme draughts forcing the fire in that particular place. It is possible to prolong the existence of a bath by carefully stopping up any leakages as they occur; and this should not be done in a hasty, slovenly or careless manner, but should be performed with the care required to attain the object in view. I have known baths effectively stopped in three or four places, and they have been worked for a considerable time afterwards. The location of a leak will soon make itself apparent by the fumes arising from the oxidation of the metal. The first step is to effectually remove all the fire and ashes from the immediate vicinity of the outbreak. This having been done, the extent or size of the leak will at once be apparent. No attempt whatever should be made to stop the leakage by thrusting the point of a poker into the hole, as the plate may have worn very thin, and such a mode of proceeding will often cause the hole to be made considerably larger and not be successful in stopping the flow of metal. If the hole is very small, only causing the metal to issue in drops at intervals, it can be stopped without lowering the metal below the hole, but if it flows in a continuous stream it is absolutely necessary to take the metal out until it is below the aperture. If the stream is large the metal should be caught in moulds while the emptying is going on, until it is lowered down to the aperture.

The next step is to have two flat bars of wrought iron, three-eighths of an inch thick and the width of the flue, inserted in the flue, one on each side of the aperture, leaving a space of 4 inches between the bars. The bars are tightly wedged, by driving iron wedges between the bars and the bath. The space at the bottom can then be filled up with small pieces of brick or stone until within a foot of the aperture. These pieces of brick having been well rammed down, some fire-clay, made into a very stiff paste, should then be put in and well rammed, so that it will afford a firm foundation, and this should reach within 6 inches of the aperture. A mixture of iron filings and sal ammoniac should now be put on the top of the fire-clay, until it is a few inches over the top of the aperture, and this also should be *well rammed* down. This should be allowed to remain a couple of hours at least, until it has gone hard, and the bath can then be filled up again. If the leakage occurs close to one of the vertical stay-bars in the flue this will answer the purpose of one of the bars.

Inside Lining for Bath.—Where a welded bath is used I venture to offer the following suggestion as an effectual means of adding considerably to the existence of a bath if it is adopted when the bath is first fixed. It is well known that the wear of the bath is altogether on the inside, except the bottom, which is effectually protected by the lead, which is always put in to a depth of 6 inches. An angle-iron welded frame is to be made the size of the inside of the top of the bath, allowing for the thickness of the $\frac{1}{8}$-inch lining. The angle-iron is to be bent with the edge outside. The lining is to be made of $\frac{1}{8}$-inch plate, of the best iron procurable, with butt joints, and the joint-plates are to be inside, and the joints riveted with flat-headed rivets, the heads to be as thin as can be procured. This lining should be made to fit the inside of the bath as close as possible, and no bottom is required, as the bottom edge will be below the lead. After it is placed in the bath the welded angle-iron rim will be placed on the top, with the flange on the top, and the back of the angle-iron to the inside of the bath. The angle-iron can be effectually fastened down by either pins and cotters through, or by clamps. To secure the bottom edge of the lining a band of flat iron, or half-round iron, would be better, made in two halves, and, when in position, secured by pins and cotters. The lining should be examined every four months, to ascertain its condition, and, when required, to be renewed; its cost bears so small a proportion to the cost of a bath, its efficiency will be at once ascertained. If only one lining should be used it must have such a material effect on the preservation of the bath that its cost will be only a small amount in proportion.

TREATMENT OF ZINC ASHES.

The name of "zinc ashes" is a trade term for what is really oxide of zinc. It is chiefly formed when the bath is not at work, and also on the exposed parts of the bath while it is at work. A large amount is formed when the bath is "let down" during the interval between Saturday and Monday, and it will accumulate largely at holiday times, if the bath is standing for a week.

A cast-iron pear-shaped open pan is required for the melting, the size about 24 inches wide at the widest part, and about 36 inches long, terminating with a lip. This is set in brickwork arranged for the flame to play under and over it. It is fired from the front, and a loose plate iron door is required. The ash having been screened in a fine-meshed riddle to remove the sand, is then placed in a tub and well washed to free it from impurities as much as possible. It is then shovelled into the furnace, but only in such moderate quantities as can be easily manipulated. When it is thoroughly hot, the spelter will begin to trickle out. It is requisite to work it about well with a small iron scraper, to coax the spelter out, which will then run into a mould placed on an iron stand just under the lip.

It is very important that no flux skimmings be suffered to mix with the zinc ashes as the fumes would be almost unbearable to the operator, and it also clogs the ashes in the furnace. When thoroughly exhausted of the spelter the remains are taken out, and after cooling, are shovelled into moderate sized casks.

In a works making a moderate output, this operation is usually undertaken by the man who attends to the galvanizing baths at night, as a part of his duties. Under no circumstances should a man be paid piecework for doing the work, as it is clear that the output is one that can easily be augmented by a man so disposed, and especially so during the quietness of the night, and the man should be one who can be thoroughly relied upon, and the output recognised as one being the result of all the attention he could spare from the baths.

These remains can now be sold as zinc ashes, to the zinc smelters and others. They are as a rule sold from a sample taken at random from the bulk, but it should be clearly defined by the seller that they are sold without any representation as to the quality, and that the sample is taken at the buyer's risk. The sale of the ashes without this precaution often leads to disputes after they have left the manufacturer's hands, and it is actually safer to sell them as they stand, and let the buyer sample them himself.

The price varies according to the price of spelter. It is sold at the works and no charge is made for casks. The payment is always stipulated for cash. It is most essential to keep the casks quite free from wet, and even if headed up they should be carefully covered up if outside, as the contents suffer if exposed to rain. If casks are not available, they can be packed in bags, a stipulation being made that the bags are returned or charged for.

Flux Skimmings.—This is the trade term for the muriate of ammonia flux when its nature as a galvanizing flux is quite exhausted. It is removed as it gets worn and dirty from the exit flux box to the inlet flux box, as it is important that the exit box be kept in good condition, or the sheets will be stained. A cheaper kind of muriate is sometimes used in the inlet flux box, as this is only requisite for keeping the surface of the metal covered, to dry the wet sheets as fast as they enter the metal. Too much care cannot be exercised in removing the spent flux from the bath, as a large amount of spelter can easily be conveyed in the flux, to the manufacturer's loss.

A wrought-iron bowl 6 inches diameter, with an iron loop handle, and the bowl perforated with $\frac{1}{4}$ inch holes, is used for this purpose. When filled it should be gently tapped on the box to shake the metal out. It is also very essential to see that the holes in the bowl be kept open. A disregard of this precaution will lead to a great waste of metal, as the operation must be repeated several times during the day. The flux skimmings are then poured into casks, which should be kept covered.

There have been several processes patented for dealing with the skimmings, but I do not consider it is worth the attention of a galvanized iron manufacturer to manipulate them. The price varies from time to time, and it is sold at manufacturer's works, casks included.

No sampling is required, but it is possible a manufacturer will reap an enhanced price after disposing of one or two lots, and when the quality is known.

Galvanizing employees—and the under-hands more especially—are as a rule very careless, and do not hesitate to drop the bucket and contents in the cask and leave it there to be poured upon by the next comer. This should be rigidly prevented, as the skimmings form into a hard solid mass, and it is apparent that the better the condition they are sent away in, it is so much more likely for them to realise the highest price obtainable.

TREATMENT OF DROSS.

The dross which falls to the bottom of the galvanizing bath and rests on the lead there, is formed from the precipitation of the spelter through its admixture with the oxide of iron from the surface of the sheets. A considerable amount is also created by the wasting away of the interior of the bath. In a few days after a bath is filled, a slight crust of dross can be felt by plunging an iron rod to the bottom of the bath. If the dross is allowed to accumulate too high, it interferes with the working of the rolls in the bath. The best plan is to clear the bath *regularly*, and this can be done within two hours on the Saturday, after the day's work is finished; and as only three men are necessary for the operation, it does not interfere with the ordinary work. To remove it from the bath, an iron "spoon" is used which should be about 18 inches by 16 inches, made of plate iron $\frac{1}{2}$ inch thick cut in an oval shape, and hollowed like a spoon. This plate should be perforated with holes $\frac{5}{8}$ inch diameter and 2 inches apart, to allow the spelter to drain through.

The plate is welded to a round iron handle $1\frac{1}{4}$ inches diameter, with a cross piece at the end. This handle should be about 8 feet 6 inches long.

The spoon is plunged into the dross, and gradually worked round the bottom of the bath until the whole of the dross is cleared out. To get the dross as free from spelter as possible, it is advisable when it is brought to the surface, to cut into it with an iron shovel which releases a large amount of spelter. The handle of the spoon should also be repeatedly struck with a hammer while the spoon is suspended over the bath. It is requisite before commencing to take the dross out, to ascertain if all the holes in the spoon are *open*. Dross is very tenacious in holding the spelter, and it requires some practice to know when the spelter has more or less escaped. At the same time, it is not by any means advisable to bring it out *too* dry. It is now emptied on to old "waster" iron (ungalvanized) sheets, placed on the floor, and when cool is carried to the dross melting house.

Before cooling, it is advisable to cut the lumps of dross into smaller portions for convenience of melting. The dross melting pots vary in size, but a useful size would be 26 inches diameter by 36 inches deep. They are made of cast iron, with sides $2\frac{1}{2}$ inches thick, and the bottom, which is round, 3 inches thick. The pot has a band projecting about $\frac{1}{2}$ inch round the top edge to strengthen it, and it rests on three legs standing about 9 inches from the ground. A circular wall of firebricks is built round, leaving a 6-inch flue round the pot, and plenty of draught holes must be left in the brickwork.

A circular cast-iron plate in two halves for the

top is advisable, as it holds the brickwork together. The tools required are a round bar $1\frac{3}{4}$ inches diameter and 5 feet long, with a loop handle; a wrought-iron bowl 9 inches diameter, and the bowl perforated with $\frac{3}{8}$-inch holes; also a bowl without holes, about 8 inches diameter. Both these bowls should be made of plate iron, the perforated one $\frac{1}{4}$ inch thick, and the other $\frac{3}{8}$ inch thick.

The moulds are of cast iron, with the manufacturer's name, initials or brand cast on the bottom. The size of plates varies, as there is no uniform size, but plates are preferred of not too large a size, say, about $8\frac{1}{2}$ inches by $5\frac{1}{2}$ inches, by 2 inches thick. The moulds should be a little deeper than the thickness of the required plate, as it is not advisable to fill them too full, as rough edges may be made in doing so.

The dross is allowed to accumulate until there is sufficient to warrant the pot being "fired up." The first heating will take 15 to 18 hours, but afterwards it will melt more quickly. In a works making a small output of sheets, it is usual for the man who attends to the galvanizing bath during the night, to attend to the dross also, he emptying and filling a pot during the night, while the fire is attended to during his absence in the daytime.

In such a case it is not necessary to force the fire, as after the first heating it will melt slowly with a low fire, and this is in fact much better than having a fierce one. When the dross is melted, there will be found some spelter floating on the top, and this should be taken off, but not to the extreme, or the dross may become *too* dry and thick to allow of it being poured into the moulds. After the spelter is taken off, it should be thoroughly well stirred up, and is then ready for emptying into the moulds. It is essential to have a cast-iron plate to lay the moulds upon, as they should lie perfectly level. Enough dross should be taken in the bowl to fill one or two of the moulds at one operation, to ensure the plates being whole, and the moulds should not be filled quite to the edge. A few particles of muriate of ammonia, scattered on the top of each plate, will smooth the surface of the plates. It is known in this form as "hard spelter," and is bought by merchants and brokers, for shipment to India, where it is used for mixing with copper and making an inferior quality of brass. It is generally bought in lots of 10 tons and upwards, and the price varies according to the supply and demand.

CURVED AND RIDGED ROOFS.

There is a considerable demand for galvanized corrugated curved and ridged roofs, for collieries, ironworks and agricultural purposes generally, and it has this advantage, that it is an outlet for a considerable quantity of galvanized corrugated sheets.

Gauges used.—No. 16 w.g. is only used where great strength is required, and No. 18 w.g. is used for first class work generally. Anything thinner than No. 22 w.g. is used only for temporary roofs. In estimating the comparative advantages and economy of the different thicknesses, it is important to remember that the thicker sheets are not only more durable, as against the weather, but also allow of the purlins, rafters and other framework being placed with greater intervals than would be possible with the thinner and weaker sheets. The flutes, or corrugations, are made of various widths, those most used in England being 3, 4 and 5 inch. Sheets with 5-inch corrugations are commonly preferred by engineers. The depth is generally one-fourth of the width, and the proportions can only be modified in the manufacture, by making special new dies. Sheets with corrugations wider than 5 inch are occasionally used where great strength is required, but in such cases only thick gauges can be used with advantage.

According to the usual plan of covering a roof with sheets of 18 and 20 w.g., the purlins are placed 6 to 8 feet apart. The exact size of the sheets of course varies according to circumstances, but seldom exceeds 20 square feet, the ordinary lengths being 6, 7 and 8 feet long, by 26, 32 and sometimes 36 inches. The sheets, when laid, overlap each other about 4 inches in the length and one corrugation at the sides, but this depends on the pitch of the roof, the overlap being greater if the pitch of the roof requires it.

Curved Roofs.—In dealing with this class of roofing, although we give the girths of roofs up to 50 feet, I am assuming that the manufacturer of the galvanized corrugated curved sheets only intends to manufacture the roofs which require the plain tie and king rods. This class of roofing, which is generally termed "self-supporting," I do not recommend beyond 25 feet wide. For wider than this, it is usual to have framed principals, varying in weight, according to the width of the roof. As this class of roofing has almost an unlimited scope, it is necessary to employ a draughtsman to make the designs.

Curving.—The curving is done with the usual curving machine, which takes either 3-inch or 5-inch rolls. The corrugator should be a man competent to do all the curving required. It is difficult to arrange a piecework price for this, as the labour entirely depends on the amount of curve required to be given to the sheets, and the same applies to the price obtained, but the latter can be safely estimated from 10s. to 15s. per ton, and this will leave a good profit.

When sheets are required to be curved of 5-inch corrugations, and the quantity is under two tons, it is usual to make a charge of ten shillings for changing the rolls specially, and the same also for changing the corrugating dies. The corrugator should either be paid day-work for the curving, or 2s. 6d. to 5s. per ton, according to the amount of labour on the sheets. And as it is not desirable to interfere with the ordinary work, I would suggest that the curving be done in the evening. A small wall engine or a vertical engine would drive the curving rolls, so that the labour of an engine driver can be dispensed with.

To obtain his template, to gauge the accuracy of his curving, the corrugator strikes out the full size of the roof, his pencil passing over a board of the required length, and the sweep is then sawn out. It is desirable to preserve these templates, stencilling the width and rise of the roof on each, and numbering them consecutively. If a register of these is kept, it will be an easy matter to find the size required, and save time in cutting another template.

For making tie-rods for curved roofs, the only plant required is a blacksmith's hearth and tools.

Girths.—The following particulars, which have been obtained with a considerable expenditure of time, can be absolutely relied upon, and will enable the manufacturer to ascertain quickly the weight of any size of curved roof from 10 feet up to 50 feet wide, with various heights from the ridge line to the crown of the roof. The span, or width, is based on "outside to outside" measurement of the wall plate, an allowance being made for eaves.

GIRTH OF CURVED ROOFS.

Width outside of Wall Plates.		Rise.		Girth.		Usual w.g.	Width outside of Wall Plates.		Rise.		Girth.		Usual w.g.
ft.	in.	ft.	in.	ft.	in.		ft.	in.	ft.	in.	ft.	in.	
10	0	1	9	11	0	22	25	0	6	0	29	2	20
10	0	2	3	12	0	22	25	0	7	0	30	6	20
10	0	3	0	13	0	22	25	0	8	0	31	9	20
11	6	2	9	13	2	22	30	0	5	0	32	7	20
12	0	4	0	15	4	22	30	0	6	0	33	6	20
13	6	3	6	16	0	22	30	0	7	0	34	7	20
14	0	4	0	16	3	22	30	0	8	0	35	10	20
15	0	3	0	16	6	22	35	0	5	0	37	0	18
15	0	4	0	17	6	22	35	0	6	0	37	11	18
15	0	4	6	18	6	22	35	0	7	0	38	8	18
15	0	5	0	19	6	22	35	0	8	0	39	11	18
17	0	5	0	20	9	22	40	0	6	0	42	10	18
18	0	3	0	19	10	22	40	0	7	0	43	8	18
18	0	4	0	20	10	22	40	0	8	0	44	7	18
18	0	5	0	22	0	22	40	0	9	0	45	7	18
20	0	4	0	22	6	20	45	0	8	0	49	1	18, 16
20	0	5	0	23	8	20	45	0	9	0	50	0	18, 16
20	0	6	0	25	0	20	45	0	10	0	51	0	18, 16
20	0	7	0	26	6	20	45	0	11	0	52	0	18, 16
22	0	5	6	25	6	20	50	0	9	0	54	6	18, 16
24	0	4	0	26	0	20	50	0	10	0	55	6	18, 16
24	0	5	0	27	7	20	50	0	11	0	56	6	18, 16
24	0	6	0	28	0	20	50	0	12	0	57	6	18, 16
25	0	5	0	28	0	20	50	0	13	0	58	6	18, 16

Fittings for Fixing. — These are, galvanized mushroom-head bolts and nuts, $\frac{3}{4}$ inch by $\frac{1}{4}$ inch; galvanized round-head rivets, $\frac{1}{2}$ inch by $\frac{1}{4}$ inch; galvanized coach screws, 3 to 4 inches by $\frac{5}{16}$ and $\frac{3}{8}$ inch; galvanized cone-head screws, 2 to 3 inch; hook bolts $3\frac{3}{4}$ inches by $\frac{5}{16}$ inch or special clips; and galvanized washers for each kind of fitting.

With curved roofs it is usual to rivet the joints (unless specified to the contrary), as it is quicker, and rivets are cheaper than bolts and nuts; but it should be stated in the quotation if riveted or bolted, to prevent any misunderstanding afterwards.

The Cost of Fittings for curved roofs, including bolts and nuts, washers, coach screws and roofing hooks, can be taken at 3*s.* per square. If rivets are used instead of bolts and nuts, 2*s.* 6*d.* per square.

For ridged roofs not so many bolts and nuts are required, as the fixing chiefly depends on the round-head wood screws, and this can be taken at 1*s.* 9*d.* per square.

FITTINGS REQUIRED PER SQUARE OF 100 SUPER. FEET.

FOR CONNECTING SHEETS TOGETHER AT THEIR JOINTS.

90 rivets ($\frac{1}{2}$ inch by $\frac{1}{4}$ inch diam.) and washers = $2\frac{1}{2}$ lbs. total; or 90 bolts ($\frac{3}{4}$ inch by $\frac{1}{4}$ inch diam.) and washers = $4\frac{1}{2}$ lbs. total.

FOR CONNECTING SHEETS TO PURLINS.

Iron Purlins.—30 hook bolts ($3\frac{3}{4}$ inches by $\frac{5}{16}$ inch diam.) and washers = $7\frac{1}{2}$ lbs. total; or 30 special clips (1 inch by $\frac{3}{16}$ inch) and bolts 1 by $\frac{1}{4}$ and washers = 20 lbs.

Timber Purlins.—30 wood screws (2½ inches by $\frac{6}{16}$ inch) and washers = 3½ lbs. total; or 30 spikes (3 inches long) and washers = 4 lbs.

Tie Rods.—The following tie and king rods will answer for the following sizes of roofs:—

Span ..	14	16	18	20	22	24	feet.
Diameter..	$\frac{3}{4}$	$\frac{7}{8}$	$\frac{7}{8}$	$\frac{7}{8}$	$\frac{7}{8}$	$\frac{7}{8}$	inch.
Weight ..	33	40	56	62	69	70	lbs. per set.

The tie rod is screwed at each end, 2 inches more than the width of the wall plate, and has two nuts and washers at each end and an eye in the centre, through which the end of the king rod goes. The king rod is screwed 3 inches at one end, with two nuts, and at the other end a flat plate, 6 inches by 2 inches by $\frac{3}{8}$ inch, is riveted on, with two holes, to bolt to the roof.

I have only dealt with roofing which does not require the aid of a draughtsman.

APPROXIMATE COST PRICES OF CURVED ROOFS.

The following Table will be found of some service as regards the *approximate* cost of galvanized, corrugated, curved, self-supporting iron roofs, of about 100 feet in length, for hay barns, corn stores, cattle shedding, and for all general agricultural and manufacturing purposes.

GALVANIZED IRON.

Width of Roof from out to out.	Price per running foot complete.	If fixed per running foot extra.	At manufacturer's works the carriage and railway fares to be added.
Feet.	s. d.	s. d.	
15	3 10	0 11	The prices are based upon roofs about 100 ft. long and upwards. Purchaser to prepare and fix the wall plates ready to receive the roof, and to lend the few poles and planks required for scaffolding. Gutters and pipes extra. Ventilator openings, every 10 or 20 ft., 20s. each extra. For small roofs about 50 ft. in length or under, add 5 per cent. to the prices.
18	4 11	1 1	
20	6 3	1 2	
25	7 6	1 4½	
30	9 0	1 8½	
35	10 9	2 0	
40	13 1	2 4½	

If ends are required they are charged extra to above prices.
The carriage and railway fares to be added to the above.

Particulars of Foregoing Prices.

The foregoing prices are based upon the following particulars: taking galvanized, corrugated, curved sheets, 10*l*. 5*s*. per ton for 20 w.g., and 10*l*. 10*s*. per ton for 22 w.g., at the manufacturers'

Width.	Gauge.	Squares.	Weight G.C. Sheets.				Weight Tie Rods.		
ft.			tons.	cwts.	qrs.	lbs.	cwts.	qrs.	lbs.
15	22	18	1	6	2	0	3	1	0
18	22	21½	1	11	3	0	5	2	0
20	20	23½	2	3	0	0	6	0	10
25	20	29	2	13	0	9	6	3	14
30	20	35	3	4	0	7	7	1	2
35	20	40	2	13	0	24	11	0	11
40	20	45½	4	3	1	4	19	3	10

I

works. Fixing at 4*s*. 6*d*. per square. Nuts and bolts, hook bolts, coach screws and washers, at 3*s*. per square. Eleven sets of tie and king rods for each size of roof, 10 feet apart, at 5*s*., 8*s*. 6*d*., 10*s*., 11*s*., 12*s*., 18*s*., 29*s*. per set respectively. The prices are for roofs about 100 feet long and upwards, and are exclusive of carriage and railway fares, to which must be added also the profit, based on the weights shown in table above.

APPROXIMATE PRICES OF STRONG CAST-IRON COLUMNS.

For supporting iron roofs, with square base, cap and web at top for steadying wall plate.

8	10	12	14	16	feet high.
13*s*.	19*s*. 6*d*.	26*s*.	33*s*. 6*d*.	52*s*.	each.

The price depends on the size (top and bottom) and the thickness of metal.

Bolts for building into foundations are extra.

TO ESTIMATE THE WEIGHT OF A ROOF.

In estimating the weight of a roof, sometimes the number of sheets of the various lengths are ascertained, and calculated at the total weight of the sheets. The following can be taken as a guide in ascertaining the weight in this manner:—

APPROXIMATE WEIGHT OF GALVANIZED CORRUGATED SHEETS.

8-3 inch or 5-5 Corrugations.	Gauge.					
	16	18	20	22	24	26
	lbs. per sheet.	lbs. per sheet.	lbs. per sheet.	lbs. per sheet.	lbs. per sheet.	lbs. per sheet.
5 feet long	33	26	20	16	14	10
6 ,,	39	31	24	20	$16\frac{1}{2}$	12
7 ,,	46	36	28	23	$19\frac{1}{4}$	14
8 ,,	52	41	32	26	22	16
9 ,,	59	47	36	29	25	18
10 ,,	65	52	40	33	28	20

These are slightly above the export standard of weights, but it is not advisable to cut the thicker gauges too fine in estimating for a lump sum price.

A Quicker Method is to ascertain the number of square feet by multiplying the girth by the length, and the following weights per square foot, or 100 square feet, include the usual allowance for the overlapping at the sides and ends.

APPROXIMATE WEIGHT OF GALVANIZED CORRUGATED SHEETS PER 100 SQUARE FEET

(Including usual allowance for overlapping).

16	18	20	22	24	26	w.g.
331	260	205	165	142	103	lbs.

Breaking the Joints.—When ascertaining the number of sheets required in a roof, it is requisite that the lengths vary, so that the joints are not in a continuous line, and this imparts greater strength and rigidity to the roof than is obtainable by the method sometimes adopted of having the joints running in a continuous line throughout the roof. A joint, also, should not come in the crown of the roof, or leakage will ensue.

For instance, a curved roof 20 feet span by 4 feet rise is 22 feet 6 inches girth. One tier should comprise one sheet each 6 feet 6 inches, 8 and 9 feet. This allows 6 inches for each of two joints, and the 8 and 9 feet bolted together carry the joint past the crown of the roof. The next tier is reversed, and this carries the joint across the crown also, and also breaks the joints, and so on to the end of the roof.

Cost of Fixing.—It is not necessary or desirable to have men permanently in employment for the fixing, as there are men who engage to do the fixing complete at the following average prices, which include everything except the railway fares. It is usual for the purchaser to provide the necessary scaffolding, and also cart the materials from the nearest railway station to the site, and this should be clearly stipulated in the quotation.

PRICE FOR FIXING.

Ridged roofs	3s. per square.
Curved ,, ...	4s. to 4s. 6d. ,,
Principals and purlins	25s. per ton.
Tie rods	30s. ,,
Columns	20s. ,,
Wood wall-plates, ridging, eaves gutters and pipes	1d. per foot.
Valley gutters	2d. ,,

Above are for roofs of 50 feet and upwards, but no doubt a concession can be made on a large job by placing a drawing before the fixer.

The fixer usually applies for a "draw" on account, as the work proceeds, to pay his underhands. It should be understood that a certificate from the purchaser of the length fixed should accompany such application, and also, before a final settlement is made, a letter should be procured, stating that the roof is fixed to the purchaser's satisfaction.

GIRTH OF RIDGED ROOFS OF USUAL SIZES.

Width out-to-out of Wall-Plate.	Rise.		Girth.		Width out-to-out of Wall-Plate.	Rise.		Girth.	
ft.	ft.	in.	ft.	in.	ft.	ft.	in.	ft.	in.
15	3	6	17	0	25	5	10	28	0
18	4	3	20	6	30	7	0	33	6
20	4	9	22	8	35	8	6	39	4
22	5	2	25	0	40	9	6	45	6

To Obtain these Orders it is requisite to advertise in the colliery and agricultural journals. Of the latter the 'Field' is decidedly the best medium, and a small illustrated advertisement in that journal will bring orders. For this class of business good prices can be obtained, and the price should contain a margin to cover the cost of advertising.

Punching Corrugated Sheets.—Some years ago a Mr. Allen, an engineer, invented a clever contrivance for the purpose of punching corrugated sheets in a systematic manner. The machine was made as follows. A cast-iron plate, formed with corrugations to receive the sheets, was made. In this plate convex holes were cast at the required distances, which were for the purpose of inserting steel dies. These dies had holes drilled in them the requisite size to be punched, to suit the rivet or bolt to be used. This bed was made 6 feet long by 10-3 inch corrugations wide. It was not a solid casting, but a frame, so to speak, say 6 inches wide at the sides and 12 inches wide at each end. A wrought-iron frame, made of flat bar iron, corrugated to suit the bed, was made and hinged to it. Through this frame were holes corresponding exactly to the holes of the dies in the cast-iron bed. The corrugated sheet was placed on the bed and the frame lowered on to it, and the holes in the frame were a guide for the punch, which drifted into the dies. The holes on the sides of the sheet were spaced about 12 inches apart, and at the end for double riveting, on alternate flutes. Any length of sheet could be punched by moving the sheet forward. The holes would all correspond, and it is much superior to the plan of using a loose template. The machine was invented principally for the *thick* gauges, as 16, 18 and 20 B.W.G. The machine was largely used on Government work in India. It is needless to say that the sheets require to be carefully corrugated. to ensure accuracy in the punching, and in the case of thick gauges they should be of good quality, *well annealed* and twice corrugated.

GALVANIZED RIDGING AND GUTTERS.

The following are the current prices, May 1, 1896:—

	Girth	12	15	18	inches.
Ridging, 6-feet lengths,	24 w.g.	–/10	–/10¾	1/1	per length.
,, ,, ,,	26 ,,	–/9	–/9½	–/11½	,,
,, ,, ,,	28 ,,	–/8	–/9	–/11	,,

GALVANIZED IRON GUTTERS.

	Size	3	4	5	6	7	8	inches.
Half-round, slip-jointed, 6 feet long		–/10	–/11	1/–	1/1	1/5	1/6	per length.
O.G. Moulded gutter, 6 feet long		–/11	1/1	1/3	1/7	1/10	..	,,

Galvanized Iron Gutters—*continued*.

	Size	3	4	5	6	7	8 inches.
Elbows for half-round gutters		1/-	1/-	1/2	1/2	1/3	1/4 each.
Elbows for moulded gutters		1/2	1/2	1/3	1/4	1/6	.. ,,

Above prices include skeleton cases, containing 100 lengths of ridging or gutters, and delivery free London or Liverpool.

COST OF MAKING GALVANIZED IRON RIDGING, PER 100 FEET.

The ordinary corrugating quality will do for this purpose, and the sheets should be rolled *light to gauge* and well annealed. The men are usually paid piece-work, and the following are fair prices for the labour. The ridging and gutter machines run by power, and one man and boy can work the machine.

24 W.G.

Girth	12 in.	15 in.	18 in.
Width of sheet to make 2 lengths	24 in.	30 in.	36 in.
	cwt. qr. lb.	cwt. qr. lb.	cwt. qr. lb.
Weight of 50 sheets, making 100 lengths	5 2 9	7 0 16	8 1 26
	£ s. d.	£ s. d.	£ s. d.
Galvanized iron, at 10s. per cwt.	2 15 10	3 11 5	4 4 10
Labour, making 100 lengths ..	0 4 2	0 4 2	0 4 2
Skeleton case	0 5 0	0 5 0	0 5 0
Cost of 100 lengths	3 5 0	4 0 7	4 14 0
= per length	7¾d.	9½d.	11¼d.

26 W.G.

Girth	12 in.			15 in.			18 in.		
	cwt.	qr.	lb.	cwt.	qr.	lb.	cwt.	qr.	lb.
Weight of 50 sheets, making 100 lengths	4	1	24	5	2	9	6	2	22
	£	s.	d.	£	s.	d.	£	s.	d.
Galvanized iron, at 11s. 6d. p.cwt.	2	11	4	3	4	2	3	17	0
Labour, making 100 lengths	0	4	2	0	4	2	0	4	2
Skeleton case	0	5	0	0	5	0	0	5	0
Cost of 100 lengths	3	0	6	3	13	4	4	6	2
= per length		$7\frac{1}{4}d.$			$8\frac{3}{4}d.$			$10\frac{1}{4}d.$	

28 W.G.

Girth	12 in.			15 in.			18 in.		
	cwt.	qr.	lb.	cwt.	qr.	lb.	cwt.	qr.	lb.
Weight of 50 sheets, making 100 lengths	4	0	2	5	0	2	6	0	2
	£	s.	d.	£	s.	d.	£	s.	d.
Galvanized iron, at 12s. per cwt.	2	8	3	3	0	3	3	12	4
Labour, making 100 lengths	0	3	0	0	3	0	0	4	2
Skeleton case	0	5	0	0	5	0	0	5	0
Cost of 100 lengths	2	16	3	3	8	3	4	1	6
= per length		$6\frac{3}{4}d.$			$8d.$			$9\frac{3}{4}d.$	

The gauges given for half-round and O G are what I consider suitable gauges for the respective size of gutter. I would advise a manufacturer to give the gauges of the gutters he quotes for when quoting for quantities. It will be noticed in the copy of current prices no gauges are stated, so that the manufacturer, in coming into competition with other manufacturers, should know if he is competing on equal ground. The agent should always be

COST OF MAKING HALF-ROUND GUTTERS.

Size	3 in.	4 in.	5 in.	6 in.	7 in.	8 in.	9 in.
Width required to be cut	6½ in.	9 in.	11 in.	12½ in.	15½ in.	17 in.	18½ in.
	26 w.g.	26 w.g.	24 w.g.	24 w.g.	22 w.g.	22 w.g.	20 w.g.
	cwt. qr. lb.	cwt. qr. lb.	cwt. qr. lb.	cwt. qr. lb.	cwt. qr. lb.	cwt. qr. lb.	cwt. qr. lb.
Weight of 100 lengths	2 3 17	3 0 26	5 0 17	5 3 6	8 3 7	9 1 14	12 3 22
	£ s. d.	£ s. d.	£ s. d.	£ s. d.	£ s. d.	£ s. d.	£ s. d.
Galvanized iron, at 10s. per cwt., 24 w.g.	1 13 5	1 17 2	2 11 6	2 18 0	4 8 1	4 13 9	6 9 6
Labour, per 100 lengths	0 16 8	0 16 8	1 0 10	1 0 10	1 5 0	1 5 0	1 13 4
Skeleton cases	0 10 0	0 10 0	0 12 0	0 12 0	0 15 0	0 15 0	1 0 0
Cost of 100 lengths	3 0 1	3 3 10	4 4 4	4 10 10	6 8 1	6 13 9	9 2 10
= per length	7¼d.	7½d.	10d.	11d.	1s. 3d.	1s. 4d.	1s. 10d.

Gutters do not lie so close in packing as ridging, and proportionately larger cases are required.

COST OF MAKING O.G. MOULDED GUTTERS.

Size	3 in.	4 in.	5 in.	6 in.	7 in.
Width required to be cut	8¼ in.	11 in.	14 in.	16½ in.	19½ in.
	26 w.g.	24 w.g.	24 w.g.	22 w.g.	22 w.g.
Weight of 100 lengths	cwt. qr. lb. 3 3 9	cwt. qr. lb. 5 0 17	cwt. qr. lb. 6 2 22	cwt. qr. lb. 9 0 4	cwt. qr. lb. 11 0 18
Galvanized iron, at 10s. per cwt., 24 w.g.	£ s. d. 2 4 1	£ s. d. 2 11 6	£ s. d. 3 7 0	£ s. d. 4 10 4	£ s. d. 5 8 9
Labour, per 100 lengths	1 0 10	1 5 0	1 5 0	1 13 4	1 13 4
Skeleton cases	0 10 0	0 10 0	0 12 0	0 15 0	1 0 0
Cost of 100 lengths	3 14 11	4 6 6	5 4 0	6 18 8	8 2 1
= per length	9d.	10¼d.	1s. ½d.	1s. 4½d.	1s. 7½d.

These gutters are bulky and for 6- and 7-inch not more than 50 lengths could be packed in a case with safety.

furnished with samples of different gauges, so that he can point out the difference to the buyer. A buyer is always anxious to get as much as he can for the money, and an exhibition of samples will often secure the order against a manufacturer who reserves the gauge in his quotation. I name this, as orders for gutters do not always come alone, and it is advisable to be well informed with this, so as to secure the order for the sheets also. Gutters are measured from centre to centre of bead.

GALVANIZED IRON PIPES.

2	2½	3	4	5	inches diameter.
1s.	1s. 2d.	1s. 3d.	1s. 4d.	1s. 8d.	per 6-ft. length.

Above are the usual sizes and the current selling prices. They are usually packed in skeleton cases, containing 50 lengths, and the cases are charged for. Two or three different sizes can be packed in a case, as they go inside each other.

The manufacturer of pipes requires a draw-bench and tools, and an experienced operator. Unless the galvanized iron manufacturer has a large demand for them, it will pay him to buy them ready packed and delivered to seaport, even if a small profit only is obtained. Unless the manufacturer has an odd-work galvanizing bath it is a hindrance to galvanize them at a sheet-bath.

Style of a Company.—I would advise anyone, whether just commencing in the trade or a manufacturer, who may have decided to roll his sheets from steel bars instead of puddled bars, not to adopt the term "steel" as a part of the style (if a Company) of the firm, or to quote for "galvanized corrugated *steel* sheets," and for this obvious reason: the term "galvanized corrugated iron" has been used for describing galvanized corrugated sheets from the commencement, and, I suppose, will be to the end of the chapter.

When galvanized there is no apparent or appreciable difference between good iron or steel sheets, as far as the appearance of the sheets is concerned, and the buyer would not be prepared to pay more for one than the other. This is a matter that only concerns the manufacturer, who knows he can ensure a fine surface by using steel, to say nothing of the economy in the manufacture of the sheets. Buyers get the sheets indented for as "galvanized corrugated iron," and a manufacturer who offers to supply galvanized corrugated *steel* sheets runs a risk of his offer not being accepted, for this reason. Buyers as a rule will not deviate, on their own responsibility, from the orders sent from abroad.

GALVANIZING ABROAD.

Several attempts have been made to introduce the industry of sheet galvanizing abroad. Having had some experience in connection with one of these ventures, I am in a position to give my opinion as to the impracticability of such an enterprise, as regards South America. A few years ago I entered into an engagement to go out to Chili to put down and start a works for this purpose. I landed at Buenos Ayres, and, after a few days' stay there, I proceeded through the Argentine by train to Mendoza, and from thence over the Andes. Part of the journey was made over the Trans-Andine Railway, as far as it was made over the Andes, and the rest of the journey on mules. After a tedious journey of five days and two nights from leaving Buenos Ayres I arrived at my destination, Valparaiso. I immediately commenced operations.

If I had foreseen the difficulties I had to encounter, I certainly should not have entered into the engagement. I found a building had been constructed of galvanized corrugated iron, which was intended for the galvanizing factory. No provision whatever had been made for escape of the fumes arising from the pickling and dipping. The roof, instead of being in one span, was actually supported or propped up with posts, about 4 inches square. I had to brace the principals, in order to remove these props, as they stood in the way of any plant being fixed. The building stood upon "made" ground, consisting wholly of sand and stones. It therefore presented a promising site for the erection of engine and boiler, corrugating machines, galvanizing baths, etc.! At the side of the factory was the bed of a mountain stream, which was dry when I arrived there, but I found after a heavy rain (and when it does rain it is indescribable) it was a raging torrent—so much so, that it poured through the factory.

I first laid down an immense block of concrete, as a foundation for the engine and boiler. After fixing this securely, I proceeded in the same way with the fixing of the corrugating machine. I never found *solid* ground, as the made-up ground was quite 20 feet deep, so that these large concrete blocks actually rested on the sand and stones. No provision had been made in the construction of the building for carrying the shafting to drive the machinery, so I had to erect strong staging for this purpose. I started the machinery without even the slipping of a belt, and was naturally pleased when this stage was reached. This part of the work took over four months to accomplish. My difficulties can be better surmised, by the fact that all the workmen were Chilian natives, who did not know a word of English, while on my part I did not know a word of Spanish, so that it was a very difficult matter to give my instructions.

I thought my difficulties were over with the successful starting of the heavy machinery, but soon found that they were only just commencing.

The duty on galvanized iron, at that time,

arriving in the Port of Valparaiso was £2 per ton, while black sheet iron did not pay any duty whatever. It was to secure this extra profit of £2 per ton, over and above the cost of manufacturing, that this enterprise was started. But alas! "the schemes of men and mice oft go astray." The outlook was certainly very promising, and it is not surprising that investors in Chili would look favourably on the enterprise, and especially so, as the Government had granted this company a concession for several years. A syndicate was formed to purchase the machinery, erect the buildings, etc., and it was expected that when a start was made the shares would be at a premium, even if procurable at all.

But to proceed with my story. While the corrugating and other machinery was being fixed, I proceeded to put in the concrete foundation for the galvanizing bath, construct the pickling tanks, etc., with a view of making a speedy start. But fresh difficulties arose at every step I took.

I may say that previous to making my engagement with the Chilian partner, who was over here, he had ordered the galvanizing bath and machinery, which was made according to *his* idea as to what they ought to be. The bath I found on its arrival in Chili was far too small, being only four feet square, and was only fit for the purpose of galvanizing buckets, and small articles. This man, with an assumption of mechanical knowledge which is characteristic of his race, had come to the conclusion, that if the sheets actually passed through the metal they would immediately take the coating. He apparently did not know, or had overlooked the fact, that a certain time was requisite for the sheet to get as hot as the metal itself, and (as it is termed) "boil off" before the coating took place. The consequence was, the bath was useless, unless the metal was heated to such a degree that the sheets emerged in all the colours of the rainbow, and at the same time caused a considerable waste of metal and ammonia.

The proprietors, acting on my advice, immediately cabled to England for baths and machinery of the proper description, and in the meantime I was instructed "to do the best I could" with the bath I had, *minus any machinery!* Any practical manufacturer can imagine my position, and my only course was to ask for my engagement to be cancelled. Previous to this I was accused of being in league with the merchants, to prevent the undertaking being a success, and threatened with imprisonment as the consequence!

After I left Valparaiso the works stood for some time, until the other machinery arrived, but this did not mend matters, as they were minus a practical man to direct operations, and the result was that after a struggle of a month or two, the affair completely collapsed.

I will now give my impressions, based on my experience, as to the practicability of carrying on such a manufacture in that country. I may say at the outset, that I do not consider it practicable if attempted by Chilians, supposing there were no other obstacles in the way. The gambling spirit is strongly developed in these people, and they carry it into their business transactions. The idea with this enterprise was to boom it, and sell it afterwards at an enhanced price. We see such things done on this side, but not so much in manufacturing enterprises; but with these people it is the rule with any business project they start. They are strictly an agricultural, and not a commercial nation, and have not the inherent spirit of manufacturing in them.

Then another barrier to success is the question of labour. There is a strong prejudice amongst Chilians as to employing English labour, and especially so in the matter of superintendence. Some few years ago a well experienced engineer was engaged as general manager of the railway system there, but he found such opposition placed in his way, he was glad to throw up his engagement, and return home.

The lack of efficient superintendence is seen in the extremely debilitated and dangerous condition of the rolling stock, which has to be seen to be actually realised. Whilst I was there, an accident occurred to the express running between Valparaiso and Santiago. The train fell down an embankment, and several people were killed. The rails were found to be uncoupled; nothing was ever said as to the origin of this, but I at once attributed it to the want of proper supervision.

I am sorry to say, English workmen do not as a rule bear a high character in Chili. The chief reason is, that a great many are addicted to intemperance, caused to a great extent by their isolation, which is not improved should they marry native women.

With the exception of myself, there was not a man on the staff of this works who understood the galvanizing process, so I was under the necessity of "breaking them in." I had to act for the whole of the processes, and a dangerous occupation it proved to be. One man in his ignorance thrust a cold pair of tongs into the metal, from which I had a narrow escape of being blinded. Another dropped a damp plate of spelter in, with a like result, so that I had to be in a constant state of watchfulness, and especially so, as the metal was being worked at a greater heat than usual in consequence of the small size of the bath. I was thankful at the end of each day if I had escaped without a serious accident. I found these native workmen apt at imitation, but deficient in originality, and without the reasoning powers of an average English workman.

After describing the difficulties of the labour

question, which is light in contrast with the question of materials, there is the question of the difficulty of supply of materials. Muriatic or sulphuric acid is not to be procured in the country, and consequently, as it was not practicable to have muriatic acid, the only resource was to import the other. It is imported in iron drums, and as it is a dangerous cargo it is carried on deck with the proviso that, in case of bad weather, it is thrown overboard. Some vessels will not carry it under any circumstances.

The impossibility of depending upon a regular supply forms another great obstacle in the establishment of this industry in South America, to say nothing of the difficulty of obtaining other materials as required.

Then, again, the sheet iron is not "sorted" at the works with the precision required for galvanizing; the consequence is a much higher percentage of waste and defective sheets than with works here.

I mentioned at the outset the great profit that was anticipated when this venture was started by saving the duty. This looked very well on paper, but it had apparently been forgotten that iron is liable both to a *fall* and *rise* in price, both acting disastrously for those who are not prepared for it, and especially so to those who being so far away are not in "touch" with the market here.

In conclusion, the final obstacles, and perhaps the greatest of all, are : (1) The galvanized iron trade is in the hands of wealthy and established merchants, who doubtless smiled when they saw this attempt to establish the manufacturing of it there. These merchants virtually command the market, and two especially are in a position at any time, if so disposed, to lower their prices to such an extent that any manufactory there would find its occupation gone. (2) The great capital required forms another obstacle. It is a very simple calculation to realise the fact that a works, to keep regularly supplied, and assuming it would take three to four months for materials to arrive, would require to have a very large stock constantly on the way, thus necessitating a threefold capital upon which a dividend would have to be made, or the affair would be useless.

I believe I have now explained sufficiently to show my readers that as a commercial scheme it is not a sound one, and it is not possible to compete with the works here.

APPENDIX.

Approximate Estimate for GALVANIZING, CORRUGATING, GUTTER AND RIDGING MACHINERY AND PLANT, *to produce* 2400 *tons Galvanized Corrugated Iron and* 100 *tons Gutters aud Ridging per annum.*

Annealing furnace and annealing pots, with roof and travelling crane. Tramway and trucks, from furnace to pickling tanks. Stone pickling and wood water tanks. Galvanizing and pickling tools. Two wrought-iron galvanizing baths (one in reserve). Brickwork and ironwork to one bath. Galvanizing machine, guides, etc. Set of chain blocks and girder. Two water boshes. Sawdust box, stove and racks. Overhead shafting, pulleys and banding. Four-h.p. engine, to drive galvanizing machinery. Cast-iron dross pot, moulds, brickwork and tools. Zinc ashes furnace, brickwork and tools. Corrugating machine, for 10-feet sheets, and one pair 3-inch dies. One pair 3-inch dies in reserve. One pair each 2 and 5-inch dies. Curving machine, with 3-inch rolls. One set 5-inch rolls. Flattening rolls. Guttering and ridge-cap machine, with a complete set of tools (if with shear blades, to attach, £25 extra). Sheet-iron slitting shears, with travelling table. Draw bench for pipes. Astragal machine. Shafting, carriages, pulleys, banding and fixing. Shafting for corrugating, etc. machinery, say, 100 feet 3 inches diameter, with carriages, couplings, driving pulleys and banding. Sixteen-h.p. combined horizontal engine and boiler. Fixing machinery. Sundries, bundling bench, weighing machine, brands, etc.

Approximate cost, £1500.

To double the foregoing production would require an additional outlay of £500.

Approximate Estimate of Cost for TANK AND CISTERN PLANT, *also the Plant for Galvanizing same.*

Punching press, for punching 7 holes at once. Travelling table for same. Angle-bending machine. Circular cutting machine, for internal holes and circular blanks. Single-ended punching machine, to punch ½-inch thick. Open-ended shearing machine. Plate-bending rolls. Smith's hearth, complete. Six riveting forges. Silent fan, or Root's blower, and piping. Shafting, driving pulleys and banding. Combined 10-h.p. vertical engine and boiler. Sundries. Wrought-iron galvanizing bath, and one in reserve. Brickwork and ironwork for one bath. Drying oven. Three pickling tanks. Travelling crane. Galvanizing tools and sundries.

Approximate cost, £500.

Approximate Estimate of Cost of MACHINERY FOR BUCKETS AND GENERAL WROUGHT-IRON HOLLOW WARE, *also Plant for Galvanizing same.*

Double-arm screw press, for bucket bottoms or keg ends. Guillotine squaring shears. Circle cutting machine. Fly press. Bending rollers. Two folding machines. Swaging machine. Two wiring machines. Two burring machines. One pair crocodile shears. Two stoves. Thirty hammers, various kinds. Three heads. One anvil. Four stakes. One anvil stake. Three hatchets. One bick iron. Three creasing irons. Three side stakes. One brazier's horse. Three mandrils. Three stock shears. Three pair snips. One grindstone. Small tools, mallets, creasers, upsets, etc. Blacksmith's hearth, anvil and block, bellows, tuyre iron, hood and tools. Wood benching and fixing. Sundries. This set of tools would be sufficient for twelve workmen. Wrought-iron galvanizing bath. Brickwork and ironwork. Pickling tank. Drying stove. Sundry tools, and fixing complete.

Approximate cost of the whole Plant, £300.

INDEX.

Advantages of South Wales, 6
Advertising, 14
Agents, incompetent, 33
"All round prices," 81
Annealing, 70
— boxes, 71
— furnaces, 71
Appendix, 134
Arrangement of works, 28
Attention to count, 23
Australian buyers, 35
— market, the, 35
A visit abroad, 15

Badly arranged works, 28
Baths for galvanizing, 93
Baylis's process, 59
"Best" and "Best best," 80
Best plan, the, 11
— process, the, 64
— way to advertise, 14
Black corrugated sheets, 87
Brands, 47
Bucket plant, cost of, 136

Cabled orders, 11
Canadian market, the, 41
Cape market, the, 40
Carasco's patent, 60

Careful packing, 23
Cash payments, 24
Chief element of success, 10
Chilian factory, a, 131
Close annealing, 70
Coating, weight of, 68
Cold rolled sheets, 84
Columns, 114
Combinations, 16
Comparative rates of carriage, 5
Competent agents, 12
Concentrated works, 9
Consignments, 17, 32
Continuous roofing, 89
Corrugated black sheets, 87
— sheets, punching of, 118
Corrugating, 49
— prices for, 79
Cost of fixing roofs, 117
— of materials, 113
— of plant, 134
— prices, galvanized iron, 56
Curved roofs, 106
— — prices of, 113

Dead charges, 32
Depreciation of baths, 94
Dipping, prices for, 79
— process, old, 62
Dross, treatment of, 103

ECONOMY in manufacture, 19
Excessive dead charges, 32
Extras on galvanized corrugated sheets, 82

FAIR price for labour, 20
False economy, 19
Felt-lined cases, 44
Fittings for fixing, 111
Fixing roofs, cost of, 117
Flat galvanized sheets, 83
Flux skimmings, 101
Foreign railway contracts, 39
Furnace, annealing, 71

GALVANIZED tinned sheets, 90
Galvanizing abroad, 125
— baths, 93
— plant, cost, 134
Girths of curved roofs, 110
— of ridged roofs, 117
Good stationery, 18
— workmen, 19
Gutters, half-round, cost, 122
— O.G., cost, 123

HALF-ROUND gutters, 119
Heathfield's patent, 57
Home trade, 39
How to fail, 24
— to succeed, 9

IMPORTANCE of packing, 46
— of the trade, 1
Incompetent agents, 33
Indian market, the, 37
Iron columns, 114

JOINT accounts, 33

LABOUR, prices for, 77
Leaky baths, 96

MANAGEMENT, 30
Manila market, the, 41
Manufacturing without knowledge, 31
Markets, requirements of, 35
Methods of packing, 42
Modern times, tendency of, 30

NATIVE workmen, 130
Night shifts, 20
— work, 19

ORDERLY works, 21
Original process, 60

PACKING, importance of, 46
— methods of, 42
— prices of, 42
Painted corrugated sheets, 87
Patent flattened, 84
— processes, 23, 34
Pickling, 73
Pipes, 124
Prices current, 12
— for flat sheets, 83
— for galvanized corrugated sheets, 80
— for labour, 77
Primary object, the, 40
Private brands, 48
Punching corrugated sheets, 118

QUALITY of felt, 44
— of spelter, 91

INDEX.

RATES for galvanized iron, 5
Remelted spelter, 92
Requirements of markets, 35
Ridge roofs, 106
Ridging, cost of, 120
Rolling from steel bars, 54
— tiles, 90

SAND *v.* ammonia, 63
Second element of success, 10
Shipments for 1894, 1
Shipping facilities of South Wales, 8
Skeleton cases, 43
South Wales, advantages of, 6
Spelter, brands of, 93
— quality of, 91
Standard of weights, 87
Stock book, 23
Stopping leaky baths, 96
Style of a Company, 124
Sulphuric acid pickling, 73

TANK-MAKING plant, cost of, 135
Tendency of modern times, 30

The best process, 64
Tie rods, 112
Tiles, roofing, 90
Trade combinations, 16
Treatment of dross, 103
— of zinc ashes, 99

UNNECESSARY waste, 31
Usual extras, 82

VALPARAISO, galvanizing in, 125
Virgin spelter, 92

WASTERS, 23
Weight of coating, 58
— of roofs, to calculate, 114
Weights of galvanized sheets, 115
What is " galvanized tinned ? " 90
— is " patent flattened ? " 84
Works badly arranged, 28

ZINC ashes, 99

LONDON: PRINTED BY WILLIAM CLOWES AND SONS, LIMITED, STAMFORD STREET
AND CHARING CROSS.

www.ingramcontent.com/pod-product-compliance
Lightning Source LLC
Chambersburg PA
CBHW022128160426
43197CB00009B/1194